CITIZEN SCIENTIST

ALSO BY MARY ELLEN HANNIBAL

*The Spine of the Continent: The Race
to Save America's Last, Best Wilderness*

Evidence of Evolution

Leaves & Pods

Good Parenting Through Your Divorce

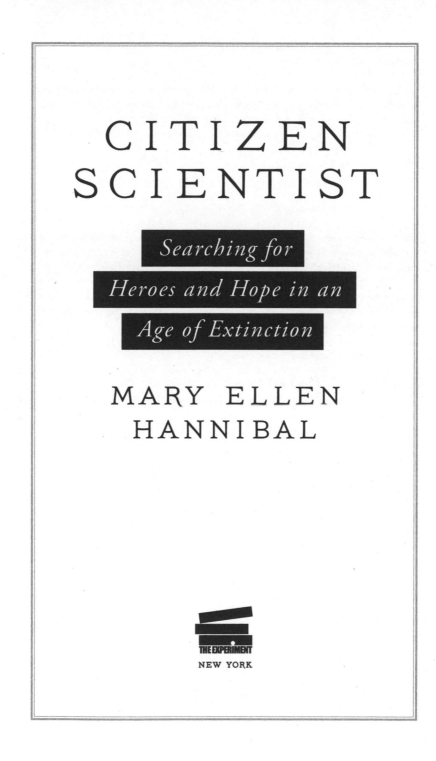

CITIZEN SCIENTIST

Searching for

Heroes and Hope in an

Age of Extinction

MARY ELLEN HANNIBAL

THE EXPERIMENT

NEW YORK

CITIZEN SCIENTIST: *Searching for Heroes and Hope in an Age of Extinction*
Copyright © 2016 by Mary Ellen Hannibal

The Experiment, LLC
220 East 23rd Street, Suite 301
New York, NY 10010-4674
www.theexperimentpublishing.com

The Experiment's books are available at special discounts when purchased in bulk for premiums and sales promotions as well as for fundraising or educational use. For details, contact us at info@theexperimentpublishing.com.

Many of the designations used by manufacturers and sellers to distinguish their products are claimed as trademarks. Where those designations appear in this book and The Experiment was aware of a trademark claim, the designations have been capitalized.

Library of Congress Cataloging-in-Publication Data

Names: Hannibal, Mary Ellen, author.
Title: Citizen scientist : searching for heroes and hope in an age of
   extinction / Mary Ellen Hannibal.
Description: New York : The Experiment, 2016. | Includes index.
Identifiers: LCCN 2016016409 (print) | LCCN 2016029859 (ebook) | ISBN
   9781615192434 (cloth) | ISBN 9781615192441 (Ebook)
Subjects: LCSH: Endemic plants--Conservation. | Endemic
   animals--Conservation. | Introduced organisms--Control. | Extinction
   (Biology) | Plants--Extinction.
Classification: LCC QK86.A1 H365 2016 (print) | LCC QK86.A1 (ebook) | DDC
   576.8/4--dc23
LC record available at https://lccn.loc.gov/2016016409

ISBN 978-1-61519-243-4
Ebook ISBN 978-1-61519-244-1

Jacket and text design by Sarah Smith
Author photograph by Richard Morgenstein

Manufactured in the United States of America
Distributed by Workman Publishing Company, Inc.
Distributed simultaneously in Canada by Thomas Allen and Son Ltd.
First printing August 2016
10 9 8 7 6 5 4 3 2 1

In Memory
Edward Leo Hannibal

You can't step in the same river twice.
—*Heraclitus*

❧

The past is never dead. It's not even past.
—*William Faulkner*

Change over Time

I was not the first to have a flow experience in the Santa Cruz Mountains. About two years ago I stood on a hillcrest above the dips and divots in the geology and looked out over the Pacific Ocean. Dr. Rob Cuthrell was talking about California Indian burning practices. Cuthrell had just received his PhD in archaeology from the University of California, Berkeley, and clearly nailed his orals—he seemed to be telling us everything he'd ever learned. I was standing with a group of citizen scientists at the location of an ancient meeting site where the Spaniard Gaspar de Portolá was given food and shelter by local Indians in 1769. Cuthrell is part of a group of scientists working alongside Amah Mutsun tribal members who trace their ancestry to the area. Together they are uncovering historical land-use practices—and working to restore them. This is not archaeology as usual.

Cuthrell is in his midtwenties, crisp and clean-cut, and that day he sparkled. Over the next couple of years, as his work at the Quiroste Valley Cultural Preserve continued, he seemed to get dustier every time I saw him. He showed us transects where the team was analyzing the density of native versus invasive plants. They had cut down Douglas fir, and the sweet smell of pine wafted in the warm sun. I scratched my head. Transects, native species? He sounded like a conservation biologist. "We can't just burn this now," Cuthrell said in his soft Southern accent. He was referring to the California Indian traditional practice of managing natural resources with fire. "We have to restore the native ecosystem first."

Cuthrell explained that the scenery, though it looked gorgeous, was full of invasive plants, some having arrived via the Spanish incursion a scant 250 years ago, others the result of subsequent ranching and dairy farming, the effects of which are still wreaking havoc though the land is now protected. "So it can't be burned until we restore the native plants," Cuthrell said. Burning the invasives could perversely enhance their populations. As I listened to Cuthrell and looked out over the landscape, I began to feel dizzy with a sense of time telescoping. The Spanish had long ago come and gone but their impact on biodiversity here was still unfolding, and consequences to indigenous people still keenly felt. We had come to this ridgetop abiding by Chronos, the ticktock by which our meeting times and dates are scheduled. But I was feeling kairos breaking into the picture. Kairos is the revelation of all time happening at once. *Now* was colliding with *All Time*.

Environmental theorist Timothy Morton has declared that "the end of the world has already occurred," dating the apocalypse: "It was April 1784, when James Watt patented the steam engine, an act that commenced the depositing of carbon in Earth's crust—namely, the inception of humanity as a geophysical force on a planetary scale."[1] The industrialization made possible by Watt is widely credited with creating climate change. Worst-case scenarios register a death knell for species as the eventual result—this bell is already gonging, because plants and animals are currently disappearing at a rate and magnitude equaling that which took out the dinosaurs.

Whether we stem what is recognized as the "sixth extinction" or not, a profound and incomparable shift has indeed occurred on earth, and the world that once was is no longer. The big five extinctions that have marked earth history over the past 3.5 billion years suggest that the fairly abrupt reduction of most species living at any particular moment is, if not a normal affair, at least one with precedence. Nature perhaps will take this one in stride. To line them up, the Ordovician, the Devonian, the Permian, the Triassic, and the Cretaceous events knocked out worlds then existing and made way for new ones. They were brought about by big climactic shifts, for the most part, just like we are experiencing today. So here's the big difference, as far as we *Homo sapiens* are concerned. The mass extinction of plants and animals currently under way is our own doing. It is the result of human impacts on earth systems. By our own actions we're threatening the very quality of life we think we're daily defending. Prior extinctions may have paved the way for humans to proliferate and dominate the worldwide ecosystem, but paradoxically our success is fueling our downfall, and it is quite plausible we are also taking ourselves out as we radically reduce the numbers of other life-forms.

The point at which humans began to have this bad effect is up for considerable debate, but looking out at the Santa Cruz Mountains I thought, "the end of the world" happened right here, when the Spanish made first contact with the Indians in 1769. Loss of native plants, animals, and indigenous lifeways reduced the efficiency of the biological carbon cycle here even before industrialization began its great transfer of carbon from ground to atmosphere. This biotic unraveling also includes a fundamental disruption of the human place in nature. I stared out at the hills and imagined the Spanish troops approaching.

On the vast ocean horizon were pointers to many of the citizen science subjects I'd been researching for this book. To my right was Pillar Point in Half Moon Bay, where I regularly monitor the tide pool with the California Academy of Sciences. For several years we have basically focused on what isn't there, because an unprecedented die-off of sea stars has wiped out virtually all of these creatures from Alaska to Baja. Closer to where I stood was Año Nuevo State Park, where I've trekked up close to northern elephant seals. The elephant seals were hunted for their oil to near extinction by the mid-1800s. Then people got into drilling for oil in the ground and spilling it all over the beaches, which is what Beach Watch volunteers are monitoring. To my left among the Santa Cruz peaks was the hilltop saved from logging in 2007 by Rebecca Moore—the gold rush assault on redwood trees in this area still goes on. Based on her work there, Moore founded Google Earth Outreach, purposing mapping and big-data capacities to empower everyone everywhere to save nature.

The work on the Santa Cruz coast falls under the rubric of a "co-created" project. The archaeologists are not taking a "me researcher, you subject" stance with the Amah Mutsun whose past they are investigating. The project has been shaped and determined equally by the PhDs and tribal representatives (some of whom also have PhDs). Co-creation is a category of *citizen science*, a term currently used to describe the widening practice of noncredentialed people taking part in scientific endeavors. Equivalent projects with indigenous people around the world are known as "extreme citizen science." What's extreme about them is that instead of subscribing to the top-down, hierarchical approach of Western science, these projects are resolutely bottom-up. They address local needs and rely on networking to create new kinds of knowledge to achieve real-world results. Co-created projects fundamentally question what science is, who gets to do it, and what it is for.

I listened to Cuthrell talk about breaking the fire bond that once deeply twined humans into their environment. Yet a full circle was being drawn here as well, a major effort to stitch together indigenous and Western knowledge systems, to restore the landscape and also to restore our sense of the human. The archaeologists wanted to quantify how people historically lived here; the Amah Mutsun wanted to restore ecological and cultural connections. Science is sometimes blamed for separating humans from nature, but here science was helping to heal the rift. Can it be healed? Are we nearing the utter end of the world, or is there a way forward? Cuthrell started to jump up and down, and for a second I literally thought some kind of cosmic completion was being revealed. Turned out he had stepped into a nest of fire ants.

The Silent History

Extinction is a word from the realm of science, but it isn't just about science. It's about history—what happened on the land and in the water, and why. History is based on storytelling, on narratives. The Spanish priests who established the missions here thought they were creating something—and they were—but they were also destroying something. They told themselves one story but they were living another one at the same time. Two things going on—and so it is today. We get in our cars and we go to work, and we work to fulfill ourselves to support ourselves and our families, and on a certain level we think, I am creating. But we are also destroying.

One of the problems with trying to grapple with this double narrative is that we have typically defined human history as separate from other categories. Historian Dipesh Chakrabarty points out that academics tend to "deny that nature could ever have history quite in the same way humans have it."[2] If we could talk to the soil, the trees, the birds flying overhead at Quiroste Valley, we would hear something else.

Humans have assumed that while our history changes in hundreds of years, the geographical environment changes only over millions of years. So there is one short history and there is one long history, and these two stories seem to run on separate tracks that only incidentally intersect. Pointing out that humankind depends on overarching narratives to make meaning and to establish power, preeminent cultural historian Lynn Hunt says that now "history's purposes are expanding as we increasingly think of ourselves as humans sharing with each other and with other species a common planetary past and future. . . . An alternative narrative is essential."[3] She might have added that we share with other species a common fate. If the biota at Quiroste could talk, we would learn that geology, biology, and human history may be investigated by us as separate chapters but, in fact, they make up one book. And the time has come for us to learn to "read" that book.

As beautiful as the view is from here, as natural as the golden hills seem, they are a patchwork of responses to human impacts over thousands of years. Human history has made ecological history and vice versa—people, other species, geology, water, and weather have made this view, and they have done it together. Ironically we know what the landscape looked like before the Spanish got here because some of them kept fastidious notes about what they saw. The answer is: wildflowers. We didn't know wildflowers lived here in magnificent abundance until about 2001, when Alan K. Brown published a new translation of Padre Juan Crespí's expedition journal from Portolá's 1769 voyage. Crespí duly documented mind-boggling fields of wildflowers, not realizing that the expedition he accompanied would fatefully suppress their blooms. Now that we know they were there, we know we can bring those flowers back. The seeds are still there, waiting for us to restore a lifeway that brought them—well, to flower and fruition.

Knowing with better accuracy what we are looking at is a virtue of citizen science. Looking down to my left I could not quite see Monterey, but I knew it was there. I could imagine Ed Ricketts trolling for specimens in the tide pools at Pacific Grove. Ricketts was a nonprofessional scientist and activist (i.e., a citizen scientist) and the model for Doc in *Cannery Row*, John Steinbeck's 1945 novel about a crew of misfits carousing among sardine canneries during the Depression era. Generations of "Ed Heads" look to approach the world as he did, on the constant search for a holistic framework. If not a bible of citizen science, *The Log from the Sea of Cortez*, his book co-created with Steinbeck, is something of a manifesto. Steinbeck and Ricketts declared they would undertake their voyage "doubly open" to objective and subjective realities. Ricketts and Steinbeck sought the "toto picture" in which art, science, and experience are integrated. What I'm trying to do in this book is what they were trying to do—put it all together, the personal, the historical, the scientific.

Like Steinbeck, the writer and mythologist Joseph Campbell was also seminally influenced by Ricketts. Campbell documented a taxonomy of the world's indigenous stories, within which he discerned a common pattern, the hero's journey. The hero's journey is universal, according to Campbell, and it unites humankind across races, cultures, geography, and time. Campbell credited an expedition he took with Ed Ricketts to Sitka, Alaska, with inspiring his understanding that the myths we live by are basically biological in origin. Myth, he said, is "nature talking."

With his emphasis on individual agency, Campbell is a special guide for the citizen scientist in today's fraught moment. Campbell provides a road map along which an individual destiny can travel to global impact. Because what is the role of the individual, when together humans constitute a geological force? Staid geologists now call our epoch the Anthropocene, because human impacts are discernable in the fossil record, in the same way as are the effects of an earthquake. Whatever we might do as separate people, our efforts are made practically irrelevant by the aggregate result of our activities. As Chakrabarty points out, this realization challenges the very notion of human history. We have thought that we were doing one thing—waging wars, claiming territories, defining human rights—but something bigger was going on at the same time.

Expeditions of the past hold keys to understanding our current terrain. Thomas Jefferson produced an expedition diary of sorts, in *Notes on the State of Virginia*. Jefferson surveys and documents the terrain, vaunting not only Virginia's natural resources but an inherent goodness he finds there. Jefferson also instructed Meriwether Lewis and William Clark in how to document the Corps of Discovery Expedition, the journey he famously sent them on to explore the western US after the Louisiana Purchase in 1803. All three of them were citizen scientists. The California Academy of Sciences was founded by amateurs. Rollo Beck's 1905 Galápagos expedition on behalf of the academy was foundational to "proving" what Darwin only posited about how life originates. The botanist Alice Eastwood saved the academy's collections in the 1906 earthquake; her lifelong collections at Mount Tamalpais in Marin County today form a baseline for understanding how climate change is impacting California vegetation. In our own time, fantastic technologies allow us all to observe with consequence. With iNaturalist on your smartphone, you can network with other nature fans and contribute real data that helps discern more accurately what is going on where, so that better decisions can be made about managing natural resources. This big-data dimension of citizen science is perhaps its most "science-y" application, relying on statistical analysis and computing power. At the same time, accurate data points on a map make it harder to misrepresent the uses and potential of a piece of land or a stretch of water. Using Google Maps, you can integrate historical and political claims with natural history and visualize that story on a map. Thomas Jefferson, citizen scientist, would have approved.

The Hero Departs

Myths are timeless, but they unfold in a setting that has everything to do with time—with the seasons, with cycles of animal movement, with the developmental moment in the hero's life. In spring 2014 when I gazed out over the Santa Cruz Mountains and imagined Portolá, I had no idea that within months, my father would quickly succumb to cancer and die before the year was out. After his death, I have continued to monitor wildlife at Mount Tamalpais in Marin County, to count invertebrates in the tide pool and hawks in the sky. The places where I do this have become ever more emotionally significant to me, even though the counting and the measuring are designed to be as impersonal as possible. The landscape and its creatures essentially contain me while I make the mysterious adjustment to before and after. Thank you, nature.

Some people like to call citizen science "participatory research." This comes out of a decades-long unfolding of thought in the humanities in which researchers began to grapple with the very unpleasant insight that they were treating their subjects as inferior objects, that it is impossible to take a "me expert, you study subject" view that is not condescending, incomplete, and more or less self-serving. With some horror, researchers looked in the mirror and saw themselves reiterating colonial control of indigenous and economically underserved people. A full-on identity crisis ensued. How are we gathering information and creating knowledge, and what are we using it for? Basically the cure for the dominator approach is to insert the word *I* into the narrative. We can't really remove bias completely, but we can state our position as honestly as possible, declare our self-interest, our subjectivity. If the researcher is also a subject, and if the subject is also a driver of the research project, then maybe we can get some equity here, and "co-create" knowledge. So here is I.

CHAPTER ONE

In Which I Freak Out
in the Tide Pool

At four in the morning on June 12, 2012, I drove down Cole Street in the Haight-Ashbury neighborhood of San Francisco, on my way to pick up a college intern carpooling with me to the California Academy of Sciences' first citizen science tide-pool-monitoring expedition. The hour ranked as dead of night, but in this neighborhood, an imperfect Age of Aquarius is perpetually dawning. I glided around a fire truck parked in the middle of the road. In the darkness, a quiet ruckus between uniforms and rags played out on a stoop.

The date marked the lowest tide of the year in these parts, brought on by the moon's position relative to the earth and the sun. This was the cosmic conjunction the intern and I were lining ourselves up with as we made a pretty straight shot twenty-five miles south to Pillar Point Harbor in Half Moon Bay, ahead of Silicon Valley traffic. My companion had a soft, ready smile even at this hour, and long blonde hair under a knit cap. I had determined our destination the old-fashioned way, by

MapQuest, and following printed-out directions we arrived at the wrong beach. The intern cheerfully typed latitude and longitude coordinates into her smartphone and directed us by degree to the right one. Kids today!

Breaking News

Citizen science is taking off as never before, and it is needed as never before. Scientists point out that while two million species have been named by science, millions more have yet to be discovered. At the same time, the aforementioned extinction crisis is taking out species before we even know they are there. What does it really take to save nature? How do we look at this gigantic problem? Citizen science starts with and continuously returns to individual observations of nature.

The hermit crabs and brine shrimp we collected over the next three days of extra-low tides at Pillar Point would help embody a snapshot in time, physical reality as it existed in this moment on planet Earth. Every day our quarry would go back to the academy, where each thing would be officially named according to the age-old methods of taxonomy and suspended in jars of ethyl alcohol. Eventually they would be accessioned, taking their place among the twenty-two-million-and-counting specimens currently housed in vast metal cabinets in a temperature-controlled basement vault in the academy's fancy building, designed by Renzo Piano, in Golden Gate Park. Thus they would join august company with specimens obscure and famous, including giant pink Galápagos iguanas brought back by Rollo Beck in 1906 and coelacanth fossils, thousands of years old, deposited at the academy in the 1970s.

Ascension!

Between the beach and the big breaking waves about a quarter mile off was a stretch of bumpy, glistening reef, its usual blanket of water pulled back by the celestial hand. The temperature was in the midforties, cold, the air wet and exhilarating, but maybe that was coffee on an empty stomach, and lack of sleep. In the parking lot I cast about for someone to follow into the drink. The academy's curator of worms (literally, that was her job) charged forth, her eagerness to find squirmy little things barely contained, reflective patches on her jacket flashing in the moonlight. The beam of her headlamp was circled by a penumbra of sea spray and then she disappeared. She entered darkness and overwhelming sound.

Treading through several inches of water sluicing gently over lumpy calcium carbonate, a.k.a. the reef, I walked in the direction I thought she'd gone, trying not to fall, and catching a sign of her now and then. It was mostly like walking straight into an enveloping dream, its essence closer at each step yet at the same time more completely obscured. One drastic wave crash after another punctuated the hurtling roll of the ultimate white noise. The constant plague of the tide pool—deep thoughts—crept in with the mist. At an edge of land and water that is only periodically revealed, I felt time at its source, unmediated through clocks and cell phones and the contrivances of our busy-ness. The seasons of course manifest the earth's changing relationship with the sun as the earth spins on its own axis and simultaneously revolves around the sun. Within this comprehensive movement, the earth is also doing a push-me-pull-you with the moon and the sun, and the resulting effect on rising and falling sea levels is known as a tide. Thus the tide pool monitoring at Half Moon Bay follows the cycle of the moon, but this operates at a scale that cannot be neatly followed by a regular schedule. We visit Pillar Point during the lowest tides that occur at hours that are reasonable for most people to be awake.

So sometimes we meet several times in a single month to count up invertebrates, and sometimes a whole month will go by with no low tides at a decent hour.

According to Joseph Campbell, the hero's journey is based on and analogous to the sun's daily round, in which both "fall cyclically into the watery abyss" and then arise again.[1] It involves a cycle of personal realization in which the nascent self heeds a call to adventure, and, following it, departs from the known and comfortable world. Trials ensue; the hero is lost. There is an encounter with a transcendent beneficial force, and the hero has a revelation that paves the way for eventual reconciliation with those he has left behind. The hero gains something—wisdom, a gift, something his people need. Newly equipped, he goes home, restoring his tribe in the process of becoming his own man. The hero's journey is literal and involves physical trials, but it has a psychological origin. Campbell referenced Carl Jung's concept of the "night sea journey," which involves diving down into unconscious darkness as a necessary prelude to rebirth. In other words, we live our lives and tell each other stories that at the molecular level are a response to this fundamental natural rhythm of sun set, sun rise.

I had a notebook in my hand and binoculars around my neck, and I stared into the darkness as if on an epic quest for . . . what? What was I looking for? The sky seemed abruptly to have had enough of my dithering and dramatically lightened up around the glowing moon, which retreated like an aging sovereign before the rising sun. It was day!

As I stood still to suss my next step, my sight was filled with a metallic melding of greens and reds and telegraphed one unified entity, the tidal reef. But the merest closer look changed everything. Distinct and dense life-forms took shape. Glowing purple sea urchins were arrayed like a wall of living burrs on three-foot shelves of reef dipping down under impossibly luxurious fronds of feather boa kelp. Gelatinous globs of sea anemones sat there with pieces of shell stuck all over them as if you had just spilled a box of cornflakes onto Jell-O molds—this was their SPF, the way they prevented sunburn when the water receded and blew their cover. Close up, colors that blended at a distance were in stark contrast, black and white, fluorescent violet, as many greens as if every pod in a field of peas were a different hue. Twelve-inch fish shifted past in surging rivulets of water. Something else moved: what, where? Everything was insanely alive, now you see it, now you don't. I thought, it's the light, it's the water, it's changing every second, it's always doing this whether I'm here to witness it or not.

For those grappling with this profusion of life and its orderly disorder, the standard orienting text is *Between Pacific Tides* by Ed Ricketts and Jack Calvin. Originally published in 1939, Ricketts and Calvin's book was the first to adequately document more than five hundred species found along the West Coast from Sitka, Alaska, to Ensenada, Mexico. The volume helps make sense of the panoply. Species are organized by their home addresses, ribbonlike zones discerned basically by how much water covers them when and how often, depending on the tide. This provides a basic lesson in ecology, because the creatures are adapted to the conditions in which they live. The urchins, for example, with their spiny grip on the reef, live in the mid-intertidal zone where the lowest tides expose them only for short periods of time. Even in semidarkness, these zones can be discerned by the citizen scientist. Stanford University Press waffled for years before publishing the book, mostly because it was revolutionary in its approach. Marine invertebrates had only ever been classified by phylogeny, or body type, an academic system hardly helpful to an initial orientation to the subject. Ricketts additionally was the first to include information about waves, tides, habitat, and predation in his inventory of critters, always presenting them as part of a holistic ecosystem.

Over the course of his life, Ricketts collected specimens from tide pools all along the coast, and many of these specimens provide a basis for comparison to help understand what we find in those places today. The academy's citizen science project at Pillar Point in a sense is paying Ricketts forward—not that he collected right here, but because participants are helping to fill in the puzzle pieces of how creatures are distributed along the coast and how those distributions have changed over time. The academy's project is but a node in a larger program run by the Bureau of Ocean Energy Management, which brings together universities; federal, state, and local agencies; and tribal governments, among others, to monitor the whole coast. Even at this level, Ricketts' influence is still felt, since his historical collection at Sitka, Alaska, provides a baseline for the northernmost site the bureau surveys.

Let Me Go

In the tide pool I was riveted by fat pink sea stars sitting like satisfied gangsters and seemingly unconcerned by their exposure; gulls would peck at them but the sea stars simply grew replacement limbs. I stared at one about a foot in diameter, with a six-inch crab stuck like a pottery shard glued by Julian Schnabel to its gullet. I was actually watching the sea star digest the crab. Later in the morning my intern friend crouched nearby. Like me, she was practically babbling with pure joy, pointing out this thing and that thing, and then, regrettably, she picked up a giant pink sea star. Only a very young person would think about physically interacting with this exaggerated form. She grinned at me, holding out her hand, draped with what I grew up calling a starfish. However, these creatures are not fish. Sea stars have an ancient lineage and strange, unique features. Their skeleton is wholly internal like our skulls, constructed out of stony, hard tissues called *stereom*. Their bizarre internal organs pump water through their bodies and move thousands of tiny tube feet for locomotion and eating. They belong in the phylum Echinodermata of the kingdom Animalia, while fish cohabit the phylum Chordata with humans and other backboned things. Plenty of scientists call sea stars starfish, so you're allowed.[2]

The intern decided to put the sea star back into the water, but it didn't want to go. It clung with all those tiny feet to her skin like Velcro drenched in superglue. She was brave while I helped pull the sea star off her and plunked it back down onto its rock, seemingly unperturbed. "I'm never going to do *that* again," she said.

Neither of us could have known it at the time, but it was possible she would never again have the opportunity to observe a giant sea star in its lair. In June 2012 our team documented approximately seventy-plus sea stars in each delineated transect. A transect is simply a measured-out plot, sometimes square, rectangular, or circular, depending on the research question and the terrain. A year later, surveys turned up five to none in some transects. Citizen science is being deployed big time to help professional monitoring operations track the tide pools and figure out the epic affiliation of twelve species of sea stars along the Pacific coast, documented from Sitka to Baja, Mexico. A sea star wasting disease is causing the biggest marine die-off yet known to human awareness.

I first talked to Dr. Peter Raimondi, chair of the Ecology and Evolutionary Biology Department at University of California, Santa Cruz, about the sea stars in March 2014, at which time he sounded fairly sanguine. After all, Raimondi had seen epidemic die-offs before. He had helped identify *Candidatus Xenohali-otis californiensis* as the cause of a "withering syndrome" that decimated black abalone in the 1990s. The bacterium attacks the abalone's gut and it stops producing digestive enzymes. Hedging against starvation, the abalone metabolizes its own body mass, which eventually shrinks the foot by which it clings to the substratum, until the mollusk cannot hang on any longer. Foothold has a special meaning for intertidal dwellers, and the abalone that can no longer moor itself is soon eaten by predators. The die-off was severe enough that the black abalone has been designated an endangered species. So far there is no evidence of recovery in any of the affected areas, which are mostly in central California.

The sea star die-off is of bigger dimensions than the abalone's—orders of magnitude bigger. While one species of abalone has nearly been vaporized, twelve species of sea star are going away fast. The abalone's range is fairly restricted, but sea stars have been observed falling apart and eventually disintegrating from Alaska to Baja. And while every denizen plays a role in the practically infinite complexity of tide pool interaction, the sea star is arguably the star of the show, figuratively as well as literally.

As unrelenting as these creatures evidently are, sea stars play a critical role in keeping things balanced in the tide pool. One of the most important ecological revelations of the past several decades was made by University of Washington ecologist and zoologist Robert Paine, who set out to quantify the effects of top predators on the rest of the ecosystem. He removed sea stars from a plot off the coast of Washington, while leaving other sea stars to their brutal devices in a control plot. The plot without hungry sea stars quickly encrusted with more and more mussel beds, until the mussels ate all the kelp, and with no kelp around, a panoply of creatures that depended on it disappeared. This top-down effect is called a trophic cascade, and parallel examples of its workings have been identified in all terrestrial and marine environments. *Trophic* is from the Greek and means "food." *Cascade* is from Latin and means "to fall." A trophic cascade operates like a domino effect, where an initial impact at the top of a structure has a direct effect on the next level of the structure and the effects keep going all the way to the bottom. Without the sea star to organize the cast, the show of life at the intersection of water and land may lose its script altogether.

The wasting had had local, seasonal outbreaks before, in 1978, 1982, and 1997, taking down sea stars in spring or summer and then literally chilling out when the water temperatures did. "We started to focus in the summer of 2013," Raimondi said, "but looking at some earlier reports, there were signs we didn't catch. In the summer, we were up in Alaska and saw the sea stars wasting away—I'd seen this before." The first account of sea star wasting came from an aquarium in Vancouver fed by fresh water. "Then it started moving south. It's marched down the coast. We thought it might be one of two things. Either it was associated with some big water issue, a local current or weather pattern, not an El Niño but something smaller. Or it could be a classic epidemiological spread over time—but that's hard to get your head around unless there are very strange currents going on in some places."

Among the potential causes that have been worried over and ruled out are pollution from plastics, ocean acidification, and radiation from the 2011 Fukushima nuclear disaster.[3] Raimondi countered these postulates, respectively, by noting that affected areas range from the pristine to the totally degraded; ocean acidification is affecting local places but as of yet has established no general pattern; and any radiated debris that might arrive at the affected areas eventually hasn't yet. Nor does he see this as a direct result of climate change, since when the outbreak began, the Pacific was in a cool phase (it has since turned very warm). On the other hand, collateral climate change effects (along with the warm water) might very well have something to do with it. Ocean currents may be shifting due to temperature changes, and general warming trends could be hampering the sea stars' immune systems. Raimondi told me the abalone die-off "turned out to be a bacteria hitchhiking on the currents, but there were no spared populations in the zone of impact. This die-off initiated by hopscotching (affecting some populations but skipping over others). This is the part that gets super mathematical and for me is super interesting—to find a pattern in something that seems patternless." By August 2013, the populations that had been skipped over in the disease hopscotch were observed to be afflicted or completely decimated.

The ocean acidification angle, however, bugged me. In February 2015 I sat on a panel in honor of a terrifying and beautiful series on ocean acidification published by the *Seattle Times*. In "Sea Change: The Pacific's Perilous Turn," reporter Craig Welch and photographer Steve Ringman told a complicated story through multilayered narratives, still photography, and video. Among other things, Welch and Ringman reported on the collapse of the oyster industry in the Pacific Northwest, where the mollusk can no longer form a hard shell. One family operation they profiled now raises their oysters in Hawaii, then ships them east.

I was seated next to Dr. Ken Caldeira of the Carnegie Institution for Science—as Welch put it, Caldeira is "the godfather of ocean acidification," which is caused by the same thing that causes warming atmospheric temperatures: ratcheting CO_2 levels. We know that sea star wasting started as higher CO_2 levels in the oceans warmed its temperatures. I turned to Caldeira on the dais. "Why can't we say ocean acidification is affecting the sea stars?" He said, "We can. Ocean acidification affects echinoderms."

By August 2014, the tally of sea stars dead and gone had reached the millions, and the disease showed no signs of abatement. The consolation of a known cause was not in sight. "It's scary," I said to Pete Raimondi. "Yeah," he admitted. "It's really, really creepy." Raimondi may not want to call the apocalypse, but since this is in fact the most extensive marine die-off yet known to contemporary history, the rest of us might at least want to call it a disaster. The term has a special resonance for the event, as *dis* is Latin for "apart," and *aster* is "star." The sea stars literally fall apart when they get this disease. Dis was a Roman god of the underworld, an association that flavors the term as if it were deity-ordained.

On the upside, sea star wasting comes at a time when we can observe and monitor it as never before. Pete Raimondi's Pacific Rocky Intertidal Monitoring Program has been surveying marine diversity for more than twenty years and incorporates data from a previous effort that goes back ten more. The program is focused on the collection of data in a uniform way by professional scientists across geographic areas, but in many places these efforts are augmented by citizen science.

"It's really a new addition for us," UC Santa Cruz research specialist Melissa Miner told me. Miner has worked with Raimondi since the inception of the rocky intertidal program. "We've thought about it for a long time. We wanted to add some aspect of citizen science to what we do, because there's huge interest in it. Some of our funders have been calling for it for a long time. But a lot of what we do requires expertise." It's hard to tell one species of sea anemone from another, and forget it when it comes to sea worms. "When the sea star wasting arose it became clear that this was a good way to involve people. On the open coast sea stars are pretty easy to identify. Sea star monitoring requires little gear and site setup is flexible."

Miner told me that Raimondi had been "a reluctant participant" in any monitoring whatsoever early on. "He came of age at a time when what you did was experiments in the field, and that was exciting. You had a question and, after not so long, some kind of answer. With monitoring, it's very long-term; you don't start with a question. You see what transpires. But now he's really come full circle. He realizes the importance of long-term data and he pushes to get it."

"Is this it?" I couldn't help but persist in asking. Miner knew what I was talking about. In July 2014, the prominent journal *Science* had produced a special issue called "Vanishing Fauna" about the accelerating rate of species extinctions. In the central piece of the issue, Stanford University environmental scientist Rodolfo Dirzo and colleagues showed how losing so many species was affecting the state of global ecology.[4] Plants and animals are not just along for the ride here—they actually create the healthy functioning of planet Earth. Then there is that persistent shadow provoked by scientific query, a potential "tipping point" in earth processes. It is possible that we will lose so many species that the way ecosystems operate will change, and not for the better. Loss of species can lead to accelerated rates of disease transmission from insects and birds, for example, to humans. Scourges like Ebola and Zika getting a faster ride on the conveyor belt between hosts. And one of a multitude of ongoing impacts on species loss could induce some kind of large-scale unraveling that would more directly threaten life as we know it.

"Is this it? Losing all these sea stars—are we about to watch something horrendous unfold?"

"A lot of people want to assign this disease event as a sign of larger-scale issues in the ocean," she said. "They're talking local extinctions, but I wouldn't conclude that yet."

"Local extinctions" refers to populations and doesn't mean the entire species everywhere goes away for good. But since "local extinctions" in this case refers to the entire West Coast of North America, the potential loss cannot be called anything less than staggering. The Pacific Rocky Intertidal Monitoring Program is now enlisting people to document juvenile sea stars. With data on how many young ones are populating the die-off areas, scientists may be able to figure out whether a recovery is under way. It sounded like good news that there are juveniles in many of the monitored tide pools. "But they are susceptible to the disease also," Miner cautioned. So we will watch and see if they grow, or disappear.

Seeing but Not Saying

In September 2014 my parents were having lunch in New York City after meeting with their financial adviser. My father felt a terrible pain in his shoulder and thought he was having a heart attack. The emergency room tests diagnosed pneumonia, and his progeny dispersed around the country had us a mutual, relieved laugh. "Money always gives him a heart attack." Late October, follow-up screens delivered the news that it was more than pneumonia. The story was: lung cancer, but he was going to be okay. "He's strong as a horse," my mother said. He had been biking ten miles a day all fall. The plan was to pummel him with radiation. "I'll get weak but then I'll get strong again, likely by late January," he told me. I flew from California to New York.

The story that he would be okay did not change—until the last minute, when it changed forever. Till just hours before he died on December 6, the story was: he's weak from the radiation and he will be getting stronger. I marveled at my own ability to go along with that story and to believe it fully, even as I also ran the other story in my head, the one that was actually happening. I had seen my mother-in-law die of cancer. I wanted to talk to my father before he started to decline from radiation. He would be a different person after treatment and I wanted a full dose of the father I knew, if not in his prime, then pretty close to it.

It takes all day to get to East Hampton from San Francisco. A close friend picked me up at Kennedy International Airport and I arrived at my parents' at about 8:00 PM. My father was sitting up in bed. With my mother we touched down on the treatment plan, the radiation under way. We talked about the melanoma he had survived decades before. "I've confronted death," he said. "I can do this."

Catching my parents up on my kids, my husband, my work, I burbled on for a few moments about the biologist E. O. Wilson. I had had a writing assignment focused on extinction and Wilson is the author of a concept key to understanding it. I told my parents how excited I was to talk to Wilson, who has exemplified the scientist engaged with social issues for many decades. From a place of happy report I quickly plummeted, as I noticed my father looking at me quizzically. He's confronting death, I thought, and here I am bringing up the ultimate end—extinction.

The next day we had our last real conversation. He had radiation in the morning and sat up at the kitchen table for lunch. "Maybe after tomorrow's treatment Mary Ellen and Carly should meet us for lunch at Bobby Van's," my father said. Carly is my niece. Bobby Van's is a legendary steakhouse that, like everything in the Hamptons, has become something it wasn't, but in the old days, it was where the writers hung out. You could see William Styron and James Jones and Willie Morris drinking together. James Salter sometimes, always looking healthier than the others. Those were the days.

"No Bobby Van's," said my mother.

We sat on the couch together. We said what the doctors said, over and over again: strength after weakness.

"The only thing I want to do," and I knew he would say this, "is finish my book." After the success of his first novel, *Chocolate Days, Popsicle Weeks*, my father wrote several more that didn't do as well. The novelist's byline, Edward Hannibal, was fated to general obscurity. Five kids. He went back to work in New York City, in advertising, and commuted home on the weekends. There was some talk of living like other people did, in the suburbs. But there is no ocean in Westchester.

On the couch I thought: I am already reviewing his life as he is still living it. He recounted the comments the Ashawagh Hall Writers' Group had about his current book project. He had been trying to find a second voice for the narrative. "I've got too much in one voice," he said. "I've finally figured it out, I know what I want to do."

On the drive to East Hampton from Kennedy, my playwright friend had told me about a similar creative hurdle. "Tracy rehearses big corrective speeches to her husband and daughters in the kitchen," I told my father, "and she says all the nasty, gangster things she's thinking but wouldn't actually say to them. She realized she could have a character do that—talk to herself, say things you would never say out loud."

"Ah," he said, appreciating this. "What you would never say out loud."

We talked about how it would be good for him to take a few months off from working on his book. Which you are not going to get to finish, I did not say.

When I was a child my father told me he recognized his calling as a writer from a young age. "I was always observing. Even while talking, living, going through every motion, I was watching myself and the situation. That's a writer. Always observing." He wrote novels on a manual Royal typewriter, in the hunt-and-peck style. Next to the tall metal typewriter on his big wooden desk sat two big piles of paper: the novel in progress, and a carbon copy of it. With unwavering discipline he ascended the attic stairs at 9:00 AM, came down for a one-hour lunch at noon, then typed until 5:00 PM. Under the eaves of the small attic room where he worked, a fly seemed perennially to dive-bomb the window whether it looked out onto bare winter branches or rioting summer green. I apprehended the act of writing through a sense of his concentration, wondering at the magic of those firm black letters marking emotion into paper. I looked at that inconceivable stack of neatly typed pages and wanted to produce my own.

Talking with him on the couch, where he sat stiffly in pain but as completely preoccupied with his work as ever, I watched the situation. He watched me watch the situation. As long as he could hold his concentration and funnel words through it, he was alive. This gaze that we were sharing, I thought, is our last mutual recognition. We observed each other observing each other. We both knew what was happening.

Losing the Audience

In February 2015 I was on my way to a regular tide-pool-monitoring expedition at Pillar Point. The low tide occurred at an early afternoon hour and the sunlight was lemony and bright. This trip had special significance for me, as it was the first time I would be visiting the tide pool since my father's death. I drove down Highway 1 from San Francisco and remembered bringing him with me on a monitoring trip the year before—I could just see his black rain boots and hear him chatting up the other citizen scientists. The memory made me pull the car over to the shoulder of the road.

Of course he loved the tide pool. He was an ocean man and lived near it in what had once been a small summer resort on eastern Long Island and was now an enclave of human glitz. But the ocean was still there. We moved to East Hampton upon his early success as a novelist, at age thirty, in 1969. The place had a literary tradition and he had an entry ticket. John Steinbeck famously lived in nearby Sag Harbor and had died just the year before. I was nine years old with four younger siblings. Later, as adults, the five of us found our own ways to the beach. In those early years we frequently piled into various station wagons and headed there. I remember bursting from the car and charging straight into hurling, hurricane-anticipating waves, fully dressed.

The Pacific crashed below me on Highway 1. It's a very cold ocean up here in Northern California—I would not jump into it without a wetsuit. I looked down over a promontory, and below the cliffs a few human forms moved slowly along the beach. The swirling waves were pounding and making the sand, and erosion was making the ground that's now soil and in which plants grow. Up from the water's edge crept all the terrestrial life-forms—I was watching creation and it was gorgeous. So the bodies die. I have never taken to the concept of reincarnation but suddenly it seemed like magical thinking to believe my father had just vanished into thin air. The whole person, the nearly eighty years of unfolding, developing personality—had that all evaporated? My father was Catholic bred in the bone and he was not afraid to die. On his deathbed he narrated his exit, in a way, and it was fascinating, he was fascinated himself with what was going on. But what *was* going on? Today, I thought, I will not report back to my father what I find in the tide pool. He isn't there anymore. I brought him news from a world beyond worlds that yet could be known, not just by identifying species but by living and narrating the experience, and he was always into it. He was no longer there to hear the story. Should I just go home?

Apertures of loss in my throat and chest were waypoints to a vanishing horizon now lacking the animating force of my father. What did my witness of the tide pool matter without his appreciative audience? I dreaded the solitude of my own observance, the anticipation that today I would see but not say. Soon, though, I got back to the real reasons for going to the tide pool, having to do with my own small effort to do the right thing. It struck me that the anemones and the crabs needed me to go look at them and to note their existence. If only to say, my father is dead but you are still here. (And to thank them for that.) Also, people were waiting for me.

It was a small group out there that day, just Alison Young from the Academy of Sciences and a couple of interns. We walked out the half a mile or so along the beach to the tide pool and then began our slow progress across the reef to our transect site. Water sluiced around my ankles and life-forms darted and squirmed. Nature could console me. Death in the tide pool is a regular affair but its life goes on. The individual is both important and not. Ecology is about relationships and I still had a relationship with the tide pools. The constant drama went on, the spawning and the hiding and the sudden eating, but it was also predictable and orderly. And consistently so beautiful, inducing, as Joseph Campbell called it, "aesthetic arrest," or the gasp-stop of transfixed attention. In *The Log from the Sea of Cortez*, Steinbeck wrote: "It is advisable to look from the tide pool to the stars and back to the tide pool again." So I would look to the tide pool again.

An intern upended a rock and giddily proclaimed she had found a young sea star. But she had found an adult in the process of disintegrating. Perhaps reflecting my mood, the tide pool seemed awfully empty not only of stars but of many other species. Few crabs, sparse sea anemones. The wild riot of color and interaction and the usual diversity of nudibranchs was reduced. Nudibranchs are psychedelic sea slugs that wave electric-colored gastrointestinal fronds in the water like drag queens with feathers, advertising to predators that they are poisonous so don't eat me. The vagaries of circumstance affect what you will see on any given day in the tide pool—the depth of the low tide, the intensity of wind, the changing light obscuring or revealing species like the occult red octopus. To discern a meaningful pattern, you need to have data collected in a uniform way over decades. I peered into the quadrants we'd drawn with tape measures around GPS points and I began to be downright creeped out, not by the absence in the transects but by an incongruous presence.

At first the sightings of *Okenia rosacea*, a candy-pink nudibranch colloquially known as Hopkin's rose, elicited delight. This sea slug looks like a piece of cake decoration. And then there was another one and another one and another one. One of our party found a coiled pink tape of *Okenia rosacea* eggs. It was as if we were watching the species proliferate under our very noses. This species shouldn't be here. It was like walking into a redwood forest and finding flamingos sitting in the branches.

"I thought *Okenia rosacea* were super rare," I said to Rebecca Johnson back at the academy in Golden Gate Park. Johnson is an invertebrate zoologist and a nudibranch expert. With Alison Young she co-directs the academy's citizen science program. Her cascading brown curls and soft, almost remote voice put one in mind of the nineteenth-century beginnings of the academy. Yet her multiple roles reflect our moment in time. Johnson coordinates and supports academy connections to local, state, and federal projects that monitor the California coast. It's worth taking a moment to call out the online community OceanSpaces (oceanspaces.org). This is an effort to network citizen science and professional monitoring of the coast, to visualize the findings at disparate yet connected places, and to provide a forum for people to talk to one another about what they are doing. Most citizen science today exists in local silos or in remote digital databases, and if we are going to save nature, we have to integrate them.

Johnson told me we were witnessing an *Okenia rosacea* "bloom," and it could possibly have to do with the anticipated El Niño weather event.

"*So* many?" I said. We frowned at each other. We were thinking and not saying the same thing.

"Have you heard about the Cassin's auklets?" Johnson asked me. Yes. This tiny seabird was dropping dead, straight out of the sky, in record numbers.

What exactly is happening is a double narrative. There is the current California drought unfolding along something of a "normal" story line. Drought is what happens here, always has. And there is a thwanging, thwacking, hard-to-pin-down perturbation to the historical weather patterns that cause drought, and that is global warming. In a sense, the finding of an out-of-range species perfectly illustrates why the academy is spearheading a citizen science initiative. As global warming and habitat loss affect nature, we have to track the pieces of the puzzle of what and how changes are occurring. Citizen science is not only about collecting data; it's about making a bridge between nature's drama and people like me. The hopeful sense is that if people like me observe what's happening up close and personal, and start to see patterns, then we will all be galvanized to do more to help nature. But at the moment I was freaking out. The coast of California includes more marine sanctuaries than any other location on earth, and still the species were disappearing. I had a very unpleasant thought, which was: Maybe citizen science can't help after all. Can it?

"Things are changing," Jaime Jahncke told me. "Something different is happening right now." Tall and deep-voiced, he hails from Peru and is director of the California Current Group of Point Blue Conservation Science. The California Current is the ocean system that more or less rules what happens to marine wildlife along West Coast shores. I have accompanied Jahncke out in the ocean, taking the water's temperature, documenting density of plankton and krill, and making other measures by which the state of the ocean is understood. Jahncke additionally heads up Whale Aware, a citizen science project that enjoins people cruising in the blustery deep to document when and where they see the cetacean giants, to help direct ship traffic away from them.

"So is the ocean melting down?" I contacted Jahncke after my own meltdown at Pillar Point. I am not a hysterical person but at some point, hysteria is actually called for. Not only were the sea stars and the Cassin's auklets dying off in unprecedented numbers, so were huge swaths of forest in the Sierra Nevada Mountains. The death of the trees was related to the drought we were undergoing, which was related to what was happening in the California Current. The Marine Mammal Center in Sausalito, the world's largest rehabilitation facility, was overwhelmed with starving and beached northern elephant seals and Pacific harbor seal pups; it rescued more in the first four months of 2015 than in the whole of 2014. On April 30 it rescued a sea lion that had crawled under a car at the corner of Marina Boulevard and

Divisadero Street in San Francisco, just blocks from where I live.

I knew Jahncke was not going to say yes, the ocean is melting down, despite his own measurements in 2014, according to which "the upper fifty meters of the ocean were two degrees Celsius warmer" than the previous highest water temperature, documented in 2008. Two degrees is a lot.

"I mean there has to be a point at which we call this thing," I said to him, appealing to the human emotion quelled but not eliminated from the heart of a scientist. "I know I know I know that yes we are in a drought and yes there is a way to actually quantify the extent to which it is due to human-caused climate change, but that there are also big ocean patterns over long periods of time that could be responsible for this, and it could have some resemblance to normal if we knew better what normal really is." I paused for breath.

Jahncke's research focus is on the California Current, which circulates clockwise off the West Coast of North America. Its impacts are so big it creates what is called the California Current Large Marine Ecosystem, which includes the terrestrial coast and watersheds.[5] The earth rotates in an eastward direction, and the water flows up to the Arctic region in the north and down to the Antarctic in the south. Wind plays its part. Blowing across the surface of the water, the top layer moves at a 45-degree angle from the wind, creating a spiral down into the water column and pulling up deep, colder water. Jahncke has told me that while what I just asserted is "correct like a textbook, the net movement of water for the complete spiral is 90 degrees," and I believe him. Whatever the angle, the spiral produces what is called "upwelling." It churns up dead and decomposing stuff from the deep, mostly detritus from decomposing plankton and other organic matter. These nutrients are put to use near the surface of the water by phytoplankton that use them to help photosynthesize, thus creating the bottom of the food chain upon which all the subsequent layers will feed. The Cassin's auklets and the marine mammals were dying in record numbers because the drought had vastly reduced the amount of krill at the bottom of the ocean food chain and they were starving. "But the sea stars don't have to do with the drought," I said to Jahncke. Their food sources were still there, but they were dying, too. "Jaime, what are we going to do?"

Quietly and in his somehow comforting deep accent, Jahncke said, "We keep doing things like helping to keep whales out of ship propellers. We keep trying to keep creatures safe and systems in place so they can help us withstand this as it comes down."

In addition to Jahncke I contacted Dr. Julia Parrish, executive director of the University of Washington's Coastal Observation and Seabird Survey Team. Parrish is a dynamo, named by the White House as a "Champion of Change," and she expressly studies the anthropogenic, or human-caused, influences on seabird health along with physical and biological measures. Her COASST project is a standard-bearer for citizen science, producing ironclad data and retaining a passionate cadre of volunteers. I asked her specifically about higher levels of CO_2 in the ocean and wondered why we can't just come out and say ocean acidification is killing sea stars.

Parrish was measured and deliberate. "Scientists don't already know everything," she told me. (Too bad!) "We can't determine all causal factors for something immediately. We measure, we compare, we test. In the case of elevated CO_2 levels, we're far more sure of the physics and chemistry—the warming, the acidification—than we are of the follow-on biological consequences. We *are* picking up biological signals," she admitted, "massive mortality events—no one denies these."

Not only has my father disappeared, life itself would seem to be disappearing. Not only does the center not hold, to paraphrase William Butler Yeats, neither does any periphery hold steady. I was momentarily envious that Yeats could see civilization ending in Irish-English conflict and that Joan Didion, who famously picked up his theme in *Slouching Towards Bethlehem*, could get depressed about Los Angeles. How pale and evanescent were their clarions of doom compared to what is going on now. If the basic materials of life are disappearing, then the violence of the postcolonial British Isles and the dull disaffection of Hollywood all get swept away into the same dustheap of inconsequence and all those human struggles are trumped by this gigantic squashing. This is not the unfolding of "endless forms most beautiful," as Darwin called evolution, but an inverse sucking away. Local extinctions lead to regional and universal extinctions. We have been told and told again that climate change impacts and habitat loss impacts have concentric waves of negative effects and this is it. Apocalypse right now.

You Can Taste It

My brother Jack's eulogy in Most Holy Trinity Church was exactly as its subject would like it if he were cognizant and not snoozing in a covered coffin at the front of the room. Jack told the packed pews about my father's versions of reality. "Once, at a party, I overheard my father tell someone a story about something that had happened to me. But he was telling it as if it had happened to him. Later, I said, 'Dad, that happened to me, it didn't happen to you.' He said, 'Oh, I know, bud. But it would take too long to set up that way. The story just works better if you tell it first person.' Another phrase that got a lot of traction in the Hannibal household was, 'Nothing spoils a story like an eyewitness.'"

I laughed with everyone as Jack told these most familiar stories. Jack touched down so lightly on my father's contradictions, he seemed to resolve them. *Nothing spoils a story like an eyewitness.* How many times had I heard that? Yet, Jack went on, "Dad had a slit-eyed, thousand-yard stare that he would level at you whenever he suspected you were lying, high on something—or worse, if he thought you had just used a word incorrectly." My father did not have a problem with maintaining these two different standards—entertain when the moment was right for it, and be honest when it counted. But of course it was all up to him to decide which moment called for what.

The Church was a major disappointment for my father. He loved going to mass, and so did my mother, especially when Father Huntington preached. But Huntington retired and one too many fire-and-brimstone sermons succeeded him. Decades of prevarication and deceit and child abuse on the part of the hierarchy repulsed my parents. Still, the basic Christian tenets stuck. Jack recounted the moral teachable moment when Pope John Paul II forgave the man who tried to assassinate him. "I asked Dad why he was doing it and Dad said because there wasn't a sin man could commit that the love of God couldn't forgive. I then asked why the man wasn't being released from jail. Dad said, 'Because you can't shoot the pope.'"

Jack talked about my father's faith, and said that Christianity provided both a context for his life and the lens through which he saw the world. I might have said the same thing even a week before he died. Jack had arrived at Southampton Hospital from North Carolina in time to say good-bye to my father, but didn't have time with him in the days leading up to it. My other siblings, my mother, my daughter, and I spelled each other keeping him company. He wanted to be read to—not the Bible, but the short stories of Ernest Hemingway. After every story he paused and discussed how Hemingway "did that," created the narrative. I read him "The Butterfly and the Tank" (one of Steinbeck's favorite Hemingway stories), and he said, "Ho ho, I'm sorry I know that now," as if exposed for the first time to the death of beauty by violence. "I read it to him yesterday," muttered my mother, walking by.

Although Hemingway was his favorite writer, there were others, like Jack Kerouac, or Anthony Burgess, whom I might have guessed he would have chosen for the deathbed. After a while Hemingway gets bleak. But other writers sound garish and inconsequential when life has indeed come down to its essence. My twenty-year-old daughter read Fitzgerald's "A Diamond as Big as the Ritz" to him and I found it excruciatingly beside the point. When she finished he said, "Fitzgerald writes about class, and that's okay. But Hemingway"—he put his hand up and framed space between thumb and index finger—"Hemingway gets to what's going on behind the words."

I had not been inside this church for years, though throughout my childhood and even on many Sunday mornings in my twenties, I sat in the pews. In memory the church is darker; they've lightened up the interior, which I grumpily do not like. The presiding priest was reasonable and authoritative. In this ritual of death, the Catholic Church shines. I remembered Father Huntington—tall, thin, white-haired, and patrician. He had been a Protestant English professor at Harvard University when one day, gazing at children on swings in Cambridge, all went black before his eyes. An indeterminate time period ensued. When his sight was restored he saw Jesus on the cross. This was one of his best stories, and sometimes it played in my mind as if it happened to me. I don't remember whether he saw Jesus superimposed on the playground, or if the vision came and went.

The quotidian world was restored. A bachelor, Huntington converted to Catholicism and became a priest.

While my own conversion experience isn't about Jesus, it is certainly connected to the creation. In *The Varieties of Religious Experience*, William James describes two kinds of conversion experiences: basically, fast and slow. He writes that they are essentially the same thing; the fast conversion seems to come out of the blue, but in fact it doesn't—the incremental promptings toward eureka are just more hidden from awareness. I'm still in the process of a green conversion, and the seeds for it were planted right here. Western science traces the roots of its first questions to Christianity, which explained through storytelling what science seeks to explain with data. Like California Indian liturgies, Christian rituals also enact a covenant with Creator, and they also acknowledge an indwelling spirit that can be felt even if not always seen in material expressions. And for all of the contradictions in the way it is practiced now and has been in the past, Christianity posits that humans have a moral responsibility toward life. As a citizen scientist I adhere to no doctrine but the laws of nature, finding like Emerson and Thoreau that mountaintops and river mouths make excellent pulpits. My observance is accomplished through counting up nudibranchs in a circle transect, and in my intention to help nature keep on keeping on, I consider myself a co-creator.

As the protocols of my father's funeral played out, my eyes drifted predictably out past an opening in the stained glass windows, where I could see the Most Holy Trinity school across the lawn. A handful of us were excused from East Hampton Middle School on Wednesday afternoons to walk here across town for Catholic doctrinal studies. We were preparing for confirmation—with baptism, communion, marriage, and death, one of the sacramental rituals of the Catholic faith. This class was most often taught by a volunteer, a mom. When I was in the sixth grade, one of these moms instructed us about the host, the small wafer congregants receive on the tongue at the crescendo of mass. "The host stands for Jesus," the mom told us. I raised my hand. "It *is* Jesus," I corrected her. I was a very obedient student and in those days sass was not the behavioral option for kids that it is today. She argued with me, which now seems ridiculous. I held my own. I didn't quote Flannery O'Connor ("Well, if it's a symbol, to hell with it") but recited other authorities—essentially, my father. She asked me to leave the room.

My parents, who despite their bohemian, groovy, we-live-in-an-artists'-colony demeanor were ardent Catholics both educated by nuns and my father by Jesuits at Boston College, were very pleased with my righteous insurrection. We had a powwow at the rectory with a junior priest (not Father Huntington) and the facts were presented. My father cleared his throat and stated that wars had been fought over this not-insignificant detail that the host is the literal body of Christ. The mom looked at me—we both knew she was going down, but her look said, do you really believe that in this flat bland cracker you get a bite of Jesus?

I had indeed worried over this profound detail for quite some time. As a child consuming the host I did not feel I was being aligned with death, resurrection, and redemption, as was being preached every Sunday. But I did have a strong sensation I got something every time that host dissolved on my tongue. At my father's funeral mass I got up and took communion, as the language for it goes. I remembered him at that meeting, leaned forward in a metal chair, mouth curled in anticipation of a good doctrinal smackdown. The young priest chuckled nervously and corrected the mom. That moment of surety felt connected to this moment of surety. The wafer melted. Jesus had long taken up imaginative residence as a historical figure for me, but at my father's funeral that sensation still arrived, of "all things visible and invisible," as the Nicene Creed has it. This sense of something more than physical things, yet apprehended through them, was part of John Steinbeck's quest in nature; in *The Log from the Sea of Cortez* he called religion an "attempt to say that man is related to the whole thing, related inextricably to all reality, known and unknowable." He commented that Jesus said this and so did Charles Darwin, and that their worldviews were "bound together by the elastic string of time."

Reading the Leaves

The sticking of some people to the creation story of Genesis to explain how life started here on earth is frequently bemoaned as a big impediment to dealing more realistically with how life-forms come and go. If life comes from a magic place and time, then we have no real responsibility toward it—and that is, indeed, a problem. But those of us who fully sign on to evolution should not just throw away the story of Genesis. In fact, the Christian explanation for creation provides the template for all subsequent Western inquiries into time, place, and purpose. The Garden of Eden and Noah's ark have starring roles in the story of life as understood by early European naturalists, many of whom tied their scientific inquiries directly into their expressions of faith—which included the belief that the spirit of God is embedded in the physical world. The Garden has a special resonance for California, since people thought they would literally find it here, and in a way they did.

You know the story. Adam and Eve transgressed God's wishes and were subsequently thrown out of the paradise called Eden, not only losing the ultimate real estate but also now subject to mortality. Adam and Eve started out in a timeless place, a heaven on earth. They ended up vulnerable to pain and death, and they received an expiration date. The idea of Eden eventually translated into something symbolic, but over quite a long period of time people took it literally and were on the hunt for the paradise of Adam and Eve. Where did it go? Eden had vanished but scripture implied it was just around a critical corner somewhere. John Prest, in *The Garden of Eden: The Botanic Garden and the Re-Creation of Paradise*, writes, "Throughout the Middle Ages the Garden was believed, somehow, to have survived the Flood, and in the great age of geographical discoveries in the fifteenth century, navigators and explorers had hopes of finding it."[6] They were unsuccessful, of course, and expanded their thinking on the subject. Maybe Eden was lurking somewhere hidden on the earth, or maybe the original creation had been scattered by the Flood. Expeditions thus began to be sent out to bring pieces of the original creation back home, where they would be reconstituted in "in a Botanic Garden, or new Garden of Eden."

For more than two thousand years the first formal European gardens were square, divided into quadrants representing the four corners of the earth. With the discovery of the New World, the four corners came to stand in for Europe, Asia, Africa, and America, as these continents became known. The great seventeenth-century gardens additionally parsed the quadrants into separate beds, each the home of a particular family of plants, and each plant had its own fastidiously determined place, which Prest likens to assigned seats at a family dinner table. If this expressed an idea of God's progeny, the gardens also reflected a belief that God's mind could be studied and known through the plant life. Laid out in figurative pages set for reference, the garden was thus an encyclopedia. Better than a book because the plants were real. These assembled reference guides revealed the many faces of Creator, and each family of plants represented a specific act of creation.

In the encyclopedic approach can be discerned the beginnings of modern science—the information collection, the careful ordering of relationships. Basically, the botanic garden was a living database. Faith and facts continued to be deeply enmeshed through the 1700s, when Carl Linnaeus developed a way to wrestle with new species brought back to Europe by the thousands in the age of exploration. In establishing the binomial system of naming—thus *Homo sapiens* or *Deppea splendens*—Linnaeus organized creation into a kind of spreadsheet, laying the ground for taxonomy and the study of evolutionary relationships, though of course he did not see his work that way. Linnaeus put the names of species in a big hierarchy, once again framing the natural world in a supernatural context of perfection as conceived by God, and headed up by God.

This all may seem very long ago and far away, but some basic formatting laid down by the botanical garden idea and by Linnaeus is still in use today, though it has been revised. Citizen scientists are frequently asked to do what research scientists do all the time in the field, which is count things up and/or measure things within the confines of a transect. Counting up hawks as they whiz by on their annual migration over the Golden Gate Bridge, my fellow citizen scientists and I take stations at the four cardinal directions and rotate once an hour. The tide pool count of nudibranchs and sea stars similarly takes place in a transect outlined as a representative microcosm of the whole reef. The transect is always a representative part of a whole, just like the original botanical gardens were microcosms of Eden.

The story of Noah's ark continues the story of the Garden of Eden—where do you think that olive branch brought by the dove came from? Linnaeus opined that the ark had landed on Mount

Ararat in Turkey and the pairs of species on it had headed out from there to the places where they were found in his day.[7] The two-by-two counting of species represents the basic scientific concept of the reproductive pair.[8] Linnaeus kept closely to idea that God's perfect mind was expressed on earth. This is actually similar to indigenous belief in a creator evident in all nature. But as the biblical Garden of Eden was a deathless place of eternal spring, Linnaeus's emphasis on "perfect" meant permanent. He felt that plants and animals were exactly suited to their environments now and forever. The problems that arose out of this idea are among the very first questions that developed into the concept we call biogeography.

Much citizen science is exploring biogeography at ever more specific scales. The basic question is: What are the conditions of life in a particular place that make it habitable by the species living there? The problem facing Linnaeus—how did creatures get to places past landscapes that would have killed them?—is faced today by many species on the move as the climate conditions they are habituated to are changing due to human impacts. One of the biggest negative consequences wrought by global warming is that it is changing seasonal timing. Spring comes earlier and winter has been virtually eliminated in parts of the world. Linnaeus thought God transcended the temporal dimension and so provided a constant wellspring by which nature lived. It turns out that time and seasons are in large part running the show.

Linnaeus is a herald of the Enlightenment, the transition in intellectual history in which thinkers and social activists sought to discern fact from fiction, to emerge from a mindset guided by inherited beliefs to a reasoned picture of things based on quantifiable observation. Aficionados of *Star Trek* may recall the 1988 episode "Datalore," which is a parable of sorts for this ongoing struggle. Data is the true-blue, real-deal twin whose evil sibling is Lore. Lore impersonates Data for bad purposes but doesn't get things quite right. Data wins in the end, and blasts Lore into another universe. This is what the Enlightenment and subsequent scientific endeavor has attempted to do—get Lore, or storytelling, out of the picture. Although Linnaeus was a devout believer in Genesis and Creation with a capital *C*, he provided a template for tracking relationships that were explainable by natural history rather than by divine fiat. Darwin's great insight that species rise from one another revises this into a genealogy, transforming a hierarchy of relationships with God at the top into a family tree. Darwin's thought was made possible by Linnaeus, even though Darwin basically did to Linnaeus what Data did to Lore.

Initiation

Darwin personifies the hero's journey, having pretty much failed to fulfill expectations of him before being given a big second chance on a surveying expedition to be undertaken by Captain Robert FitzRoy of the HMS *Beagle*. Darwin's job was to keep the respected but depressive leader company. He was encouraged to collect and observe natural history, though the *Beagle* had an official naturalist. For Darwin, the trip was undertaken in the spirit of a *Wanderjahr*, that tradition in which mostly young men of a certain (upper) class travel on either end of a formal education to supplement book learning with lived experience. Joseph Campbell later extolled such interludes as pivotal in the making of the hero's life. We call it a gap year. Inspired by Prussian naturalist and explorer Alexander von Humboldt's *Personal Narrative*, Darwin kept a diary of the expedition. Slated to be gone for a generous two years, the ship embarked from Plymouth Sound in the English Channel in 1831 and did not return until 1836.

The avowed purpose of the *Beagle*'s trip around the world was to map the South American coast, determining longitudes and thus improving navigation in the service of those twin objectives of empire: warfare and commerce. (FitzRoy's resulting maps were used until World War II.) Darwin had no real duties aboard the ship, and he suffered from horrible seasickness. Years later he offered prescriptive advice for the burgeoning scientist in a chapter on geology in the British Navy's *Manual of Scientific Enquiry*. He exhorted the aspirant to read, to observe, and to compare. And indeed, Darwin read, he observed, and he compared what he had seen and heard to develop the text that would eventually be called *The Voyage of the Beagle*.[9]

Darwin's reading aboard the *Beagle* included the first volume of Charles Lyell's *Principles of Geology*, and the impact of this work on his thinking was like the subject, seismic. One of Lyell's main points is that the processes that made Earth are yet in present motion. There is a constant interplay between internal heat forcing the uplift of mountains and then weather eroding them down. This has gone on through time and continues to go on. Lyell supported the contention of James Hutton, a Scottish farmer who first proposed this idea, known as uniformitarianism in contrast to the catastrophism exemplified by Noah's Flood. Lyell held to an idea of "centers of creation" to explain how different assemblages of plants and animals were to be found in different parts of the globe, and Darwin dutifully kept his eyes open looking for such centers. While he was in Chile, the usually invisible motion of earth processes took a diva turn in the forefront drama of a major earthquake, and Darwin experienced firsthand the forces asserted by Lyell.

My copy of *The Voyage of the Beagle* is collected in a single volume called *From So Simple a Beginning*, in which E. O. Wilson has assembled and written introductions to Darwin's big four, also including *On the Origin of Species*; *The Descent of Man, and Selection in Relation to Sex*; and *The Expression of the Emotions in Man and Animals*.[10] Of its total of more than 1,700 pages, Darwin's Galápagos chapter in *The Voyage* is only twenty-four pages long. It is a catalog of eye-opening wonder and appreciation. It is equally literary as it is scientific. John Steinbeck and Ed Ricketts looked to *The Voyage* for inspiration in conceiving *The Log from the Sea of Cortez*, which they would produce about one hundred years later. Darwin, Steinbeck wrote, "wanted to see everything, rocks and flora and fauna; marine and terrestrial." This is in contrast to the general scientific approach of drilling down to a single study subject becoming common by Steinbeck and Ricketts' time. "Out of long long consideration of the parts," Steinbeck wrote of Darwin, "he emerged with a sense of the whole."[11]

He didn't know it at the time, but in *The Voyage of the Beagle* Darwin was providing a prequel to his eventual book *On the Origin of Species*. His language is from the world of storytelling rather than that of science, and his details about the life-forms he sees often have a gothic tone. Upon reaching the Galápagos, Darwin reported spending the night on Chatham Island (an older name for San Cristóbal Island), of which he observed, "The entire surface of this part of the island seems to have been permeated, like a sieve, by the subterranean vapours." "The day was glowing hot," he reported, and ambulating over the inhospitable landscape "very fatiguing; but I was well repaid by the strange Cyclopean scene. As I was walking along I met two large tortoises. . . . These huge reptiles, surrounded by the black lava, the leafless shrubs, and large cacti, seemed to my fancy like some antediluvian animals." The Cyclops is from the realm of myth and "antediluvian" (before the Flood) refers to Noah. Darwin thus situated his Galápagos observations in a timeless prehistory, with reference to the Christian creation story itself.

Darwin was pretty much a creationist while aboard the *Beagle*, but in the years after he got home he cogitated on what he had seen at the Galápagos and credited the place with instigating his theory of evolution by natural selection. He revised the raw materials and then the published versions of *The Voyage*, each subsequent edition more deliberately laying out the pathway to his eventual synthesis in *On the Origin of Species*. At the Galápagos he saw strange and wondrous creatures but he also observed time in operation the way Lyell described it. As a storyteller does, he put the interactions between surface and depth—in his case, volcanic rock and oceanic currents—in a temporal framework, "once upon a time." Reading the rocks, Darwin noticed the older Galápagos islands had eroded enough to find vegetation developing on them, while the younger islands were still "covered with immense streams of black naked lava," with nothing yet able to find purchase for growth.[12] He connected the long story of the past with the short story of what was happening now. Volcanoes and earthquakes—not God—made new centers of creation upon which new species were found. The new species were related to those nearby on the South American mainland, and their differences marked the intersection of past and present.

Similar-but-different is the initial setup for speciation, which can occur when populations are isolated from others of their own kind. This is the part where new characters come on stage. Darwin saw that the volcanic islands must have recently punctuated "the unbroken ocean. . . . Hence, both in space and time, we seem to be brought somewhat near to that great fact—that mystery of mysteries—the first appearance of new beings on this earth."[13] Darwin the collector provided an ongoing inventory of his catches: twenty-six land birds, eleven "kinds" of waders and waterbirds, twenty-five species of beetle, and so forth. He updated the reader on the analysis of the ornithologist John Gould, to whom he entrusted his avian quarry, and on the progress of the botanist Joseph Dalton Hooker, who wrote up a flora of the islands. Both Gould and Hooker worked far from the Galápagos, back in England, on specimens sent and brought back by the *Beagle*.

In *The Voyage*, Darwin said it took him a while to appreciate that "the different islands to a considerable extent are inhabited by a different set of beings." He realized this was so when the vice-governor of the Galápagos told him he could tell which island a given tortoise hailed from just by looking at it. "I did not for some time pay sufficient attention to this statement, and I had already partially mingled together the collections from two of the islands." The "specimens of the finch tribe" were also unfortunately mixed up by Darwin before he fully got that the key to figuring out what makes a species so lies in knowing exactly where it comes from. If Darwin had had a smartphone and the iNaturalist app (to be discussed in greater detail later), he could have emailed photographs of the finches to Gould. Gould wouldn't have even asked which island each bird came from, since iNaturalist would locate them instantly in both space and time, by means of latitude, longitude, and clock.

No Species Is an Island

Most of the scientists signing on to conservation biology, from whence springs citizen science, came from a branch of evolutionary science focused on populations. In this they were, and are, Darwin's heirs. How could populations of plants and animals on the islands be similar to but not the same as many corollary species on the Ecuadoran mainland? The British naturalist Alfred Russel Wallace was pondering populations on islands at the same time Darwin was, and the two of them came to the same conclusion. The volcanic islands had emerged with no species on them, and gradually, immigrants from the mainland (and other places) colonized them. Darwin and Wallace saw that extinction must play its part in the establishment of new life, by taking out the old and thus creating space for newcomers. This is where characters exit the play and don't return for a curtain call. Over the long time frame, the interaction between immigration and extinction basically establishes island populations in the numbers and places we find them. This is the constant, real-time drama of life. We no longer live in that sort of real time—ours is turbocharged.

We call the theory of evolution by natural selection Darwinism, but as science historian Janet Browne suggests, we might be calling it Wallacism, if Alfred Russel had been more assertive about getting credit for his observations, and also if he had come from a higher social class.[14] Wallace is a kind of poor country cousin to Darwin's wealthy leisured gentleman, but the financial necessity Wallace felt and Darwin did not had a positive outcome for the field of biogeography. After his famous trip to the Galápagos, Darwin more or less spent the rest of his days cogitating in his backyard, while Wallace passed years in far-flung destinations the world over. Darwin exemplified the genteel nineteenth-century naturalist and is a citizen scientist exemplar. He built his monumental case for natural selection based on years of painstaking observation, under the aegis of no institution, and he did not have an advanced academic degree. Wallace is also a model citizen scientist. He had no formal education at all. Like Ed Ricketts, he made money by collecting specimens and selling them. Having worked as a mapmaker and a surveyor, Wallace prefigures the way much biodiversity science gets done today.

Traveling the globe mostly as a freelance collector, bringing back specimens to wealthy sponsors at home in England,

Wallace was, as Darwin was also, an inveterate observer, and he documented who lived where. Tallying up the geographic distribution of animals, eventually he discerned six regions spanning the globe with assemblages of flora and fauna that were distinct from one another. Although his contribution to the theory of evolution remains relatively obscure, the term *Wallace's Line* still sticks,[15] to describe what science writer Jared Diamond calls "the sharpest and most famous boundary in the world."[16] Wallace's Line divides the Pacific and Asiatic faunas, but each of the six sites has equally distinct residents. Diamond says, "Early European naturalists...were astonished when sixteenth-century explorers of other continents began bringing home exotic hummingbirds and lemurs rather than animals from familiar European groups. Why don't climatically and structurally similar habitats on similar continents support similar species?" Since God must be up to something here, it was proposed that these areas represented six different Edens, where God had "exercised creative imagination differently at each site."

Wallace figured it out otherwise though, discerning that land bridges had appeared and disappeared over periods of increased glaciation throughout the long history of Earth. Animals and plants that were once connected to each other became isolated and, with time, distinct from one another. Wallace's book on the subject, *The Geographical Distribution of Animals* (in which he also grappled with the extinction of the Pleistocene megafauna), and its sequel, *Island Life*, both made big contributions in the unfolding of evidence to support the natural selection idea and also laid the groundwork for the formalization of biogeography as a science rather than a religion.

The Art of Losing

Plans were changed. I headed to New York to take the helm of

Thanksgiving, arriving a few days ahead of other family members set to convene. My parents had been intending to spend the holiday with my sister Ellie, in Richmond, Virginia, where her household included young twins. "I want to go to Virginia," my father told my mother, upon getting the cancer diagnosis. Since travel was out of the question for him, Ellie and company would come north. "I guess this is the end of the deliveries," my mother said. Ellie had been sending See's chocolates every day and the piled-up boxes in my father's office had developed a presence. "She'll bring more," I said.

My brother Eddie and his wife, Natalie, are chef-owners of the Glass Onion Restaurant in Weaverville, North Carolina. They are open for Thanksgiving so Eddie would come the day after. My brother Jack and his wife, Luciana, were in the process of moving from Los Angeles to North Carolina. Since the story of my father's illness was weakness before strength, and since we had collectively decided my mother needed company and help, we made a schedule of coverage. Eddie and I would overlap one day and he would take over my duties for ten days. Then once Jack was settled in his new home, he would come for ten days. Then Ellie would somehow get enough babysitting coverage and come back. My sister Julie, who lives several towns over, would end up doing the most because she would fill in all the interstices of coverage. Our schedule extended through March, when strength was promised to return with the robin. I planned to come back in spring to enjoy the renaissance of vitality all around.

Ellie rolled in with her twins and her affable husband, Neil, and Julie came over with her daughter, and the small house started to rock and roll. Ellie and Julie read to my father, and the twins made brief appearances in his room. "I don't want to freak them out," my father said, looking stiff and slightly bloated up on his pillows. One of the twins crawled onto the bed and curled up next to him. The child intuited the heart of the matter, allowing us silent recognition and even joy in this. My father's delight in the twins equaled the wide-open hilarity of a daily delivery of chocolates.

All was well. My sisters read to my father from the Finca Vigía edition of Ernest Hemingway short stories. This volume strings the stories together without any information about how they were originally collected. For example, it annoyingly dispenses with any explanation for why there are vignettes about bullfighting in front of the stories from *In Our Time*. But my father knew his Hemingway and he loved these vignettes. The more bloody and direct they were, the better. "When he started to kill it was all in the same rush," wafted from the sickroom into the hallway. "Ho ho ho, that's a great line!" he bellowed.

Downstairs I cooked and cleaned and didn't sit as much with my father as I had before people started to arrive. There were a lot of bodies in the room. I realized I was replicating a historic family pattern. I'm the oldest. I got my father to myself before my siblings came along and never liked being part of the big circus of zinging interactions my father loved being the center of. Leaving the room had long been my MO. I should go back upstairs, I thought, this is no time for reversion to type.

My daughter arrived from college in Massachusetts on her Thanksgiving break. She sat with him. "Poppy asked me if I have any attachments," she told me. I rolled my eyes. She was twenty years old, of course she had "attachments"! "He said he doesn't have any attachments," she reported, and my stomach fibrillated. My father never had a single intimation of Zen. He was not one for equanimity. He tended to outbursts and reconciliations. What was he talking about "no attachments" for?

We were all living a moment of which we could not quite take the measure. We were all observing ourselves and watching my father watch us, and we were reading Ernest Hemingway short stories out loud all day long. There was a swirl of telegraphed messages as yet not cohered. We were baking and cooking and preparing. My daughter, Eva, said, "This is like a Virginia Woolf novel. *Mrs. Dalloway.*" "What about Hemingway?" I asked her. She commented that Hemingway's voice in the house was like the theme of war threading through the party-making day in *Mrs. Dalloway.* Alone with him I reported this observation to my father who beamed with pleasure. He liked for life to line up with literature.

Eva and I left on Sunday for New York City, where I hoped we could have a bit of fun before she headed back to school and I headed back to my husband and son in California. As much as I wanted her to soak up time with her grandfather, the stressful atmosphere seemed a bit much for someone who was shortly to face microeconomics again. I wanted her to have a break. In allegiance to the narrative of "weak now, strong later," we cheerily said good-bye to my father.

"We're going into the city, Dad," I said.

He had gotten progressively more immobile and he made a scowling face.

"I'm coming back," I said.

"When?"

"In a few months," I replied, "or maybe earlier."

"That's too soon!" he laughed. The bastard. I laughed too.

Eva stepped forward. "Good-bye, Poppy, I'm heading back to school." His eyes filled with tears and he averted his face. He would not look at her. Why wouldn't he look at Eva? *He knows he will never see her again.* Even as he was doing it, he couldn't bear to say good-bye to the future. I registered this somewhere in my body but silenced the thought. Time stands still. We don't know when the future ends.

Waiting at Kennedy for a delayed flight home days later, I got a phone call from my brother Eddie, who had taken over the majordomo role from me as planned. My father was at Southampton Hospital with pneumonia. This was good news, because now the pneumonia would be treated and it would stop interfering with his recovery from the radiation.

"Great," I said.

In everyone's life there comes a time to sob in an airport terminal and this was mine. I folded into a plastic seat like disintegrating paper. We were just going to hold on to this narrative until the end! But I knew the ending. Tears consumed me. The dividing line making a double narrative of what was going on was there no longer. I would not see or talk to my father again. People looked at me and did not look at me at the same time. A child stared for a long moment but then was not interested. I was grateful to this instinctive ability to acknowledge and to not acknowledge. I was not fooled by grief's advance scout. The full invasion was on, beginning already to redraw the map of my life.

CHAPTER TWO

Moby Ghost

Fall 2013. Having groped through the dark boat slips of pre-dawn Sausalito to find the cheery-seeming *Fulmar* docked and awaiting its research mission, I was happy enough to board and stand on its deck as we chugged under the Golden Gate Bridge headed out toward the ocean. Not every citizen science expedition goes out on a boat, but very significant precursors to the practice have: the seventeenth-century Spanish expeditions, for example; the eighteenth-century voyages of Captain James Cook; the nineteenth-century trips of Alexander von Humboldt and Charles Darwin; and the early twentieth-century foray of John Steinbeck and Ed Ricketts. Every citizen science trip that goes to collect evidence of what's out there takes its place in this historical timeline of exploration.

Aboard the *Fulmar*, it didn't seem quite polite to talk yet. Private sleep lingered in the faces of my fellow passengers. White flecks of reflected dawn folded and fluttered in a vast curtain up and around the bridge suspensions. Were these birds always here at this hour? Having traveled hours to see such congregations in the past, I was a bit abashed at this incredible abundance so close to my home. If I just got up real early, could I see this all the time?

"It's the anchovies," said Dru Devlin from the Farallones Marine Sanctuary Association, reading my mind.

"Oh, *food*," I said. "Of course."

With Point Blue Conservation Science—a research nonprofit employing approximately 140 scientists—the sanctuary makes three to four annual trips of about six days each to measure and document what's going on with whales, birds, and sea mammals around the Farallon Islands, a national wildlife refuge thirty miles from the Golden Gate Bridge and twenty miles south of Point Reyes. The Farallones share a zip code with the Richmond neighborhood of San Francisco (94122) but are the center of a major Pacific food web that brings birds on the wing from Hawaii, great white sharks from the middle of the Pacific, and hundreds of whales from Mexico to feed, among many other species following the same calorie trail.

Food was to be a memorable feature of my experience aboard the *Fulmar*. I had been advised that to try to prevent seasickness, I ought to take one Bonine before I went to bed the night before, to let it percolate into my bloodstream, and then to take another upon rising, at least an hour before getting on the boat. This particular day's trip had been postponed twice due to weather. Luckily, in part due to steady ingestion of junk food, I was spared feeling anything worse than a frothy sugar high, which was really pretty fun. But a graduate student who had been just fine on the several previous days out at sea took a poor turn. She executed her duties quickly, with her head down, and between tasks sat very, very still in a chair on deck at some remove from everyone else. We tacitly gave her wide berth.

Inviting bowls of Goldfish crackers, saltines, and M&M's were stationed in various nooks aboard the *Fulmar*.

"I understand the crackers," I said, grateful for the stomach-steadying support. "But M&M's?"

"They have to be peanut," Devlin said with straight-faced scientific surety, "or we won't see any whales." She advised me to keep my stomach full of them all day long. Check!

The Watery Abyss

The duties deployed aboard the *Fulmar* that day included the test-driving of a new citizen science app called Spotter Pro, the purpose of which is to help keep ships and boats from colliding with whales in the water. But under the direction of Dr. Jaime Jahncke, California Current director of Point Blue, there was a lot more going on that day in addition to Spotter Pro. As his Point Blue bio explains, Jahncke studies "spatial and temporal relationships between oceanographic processes, zooplankton, and marine birds and mammals" in the Gulf of the Farallones to "better understand food web dynamics, identify predictable hotspots, and improve ocean zoning." Thus the researchers aboard would be counting up critters both above and below, documenting water temperature and density of krill at various depths, and so forth.

Among those working for Jahncke on board were Devlin, Kirsten Lindquist, and Jan Roletto, who help run a major citizen science coastal-monitoring platform called Beach Watch. Today their work was more on the science side of things. Jahncke would synthesize and analyze the findings from this expedition and others, steadily revising a picture of what is going on in the ocean that would then inform how Devlin, Lindquist, and Roletto direct their volunteers. Jahncke might point them to specific questions to refine their data. With what they learned from him, the Beach Watch triumvirate would refine their own picture of what's going on out there, which they in turn would share with their citizen workers.

The world of ocean and coastal monitoring is confusing, which given the sheer breadth of its study subject, is perhaps understandable. As I kept tabs on Beach Watch and Point Blue citizen and professional attentions on ocean conditions between 2012 and 2016, I developed a sense of how the two camps complement each other and form a network of information. Jahncke essentially analyzes what's going on in the water to feed bird, mammal, and fish populations, or not. On the Farallon Islands, Point Blue scientists who live there for weeks at a time and have collectively occupied them for decades keep tabs on seabird breeding activity and compare this data with Jahncke's. The Beach Watch volunteers monitor the other end of things. In 2014 they documented the surging number of Cassin's auklet corpses along the shore. In 2015, an unprecedented die-off shifted attention from the krill-eating auklet to a massive mortality of the fish-eating common murre, with almost two thousand birds dead in the month of September alone. All of which was perhaps the result of an encroaching El Niño event enflamed by the effects of climate change. The rocky intertidal monitoring headed up by Pete Raimondi is yet another major component of this very large grapple with a very large subject.

"The citizen scientists capture big-picture data," Lindquist explained to me. "Then we can go out with Jaime on the *Fulmar* and get denser data points. Fishermen with the whale-spotter app in their hands also contribute to the building picture of what's going on." In recent years that picture has been mightily perplexing. In addition to the historical mortalities of some seabirds and sea mammals and sea stars, humpback whales have been coming so close to shore I have been able to watch them from Hawk Hill in Sausalito and Land's End in San Francisco. On the *Fulmar* we would see three humpback (and two blue) whales the whole day, far out into the ocean. In 2015 I went on a commercial "fast raft" whale watching trip, and at one point we were surrounded by at least forty whales, breaching, spouting, and splashing just off the shore of Monterey. The sightings don't reflect changes in the population numbers of the whales, but the fact that their fishy food supply has moved so close to land.

Scientists don't know what to make of this yet—but the shifting ocean patterns are part of what Jahncke asserted when he told me, "Things are different now. Things are changing." Nobody knows if these changes are entirely bad for all species—though they certainly are for some—or the degree to which human-driven change is responsible relative to historical weather patterns and their impacts on ocean currents. It is hard to resist the narrative that ocean warming and overfishing are impoverishing the ocean's food supply for its own inhabitants, thus speeding the course of disease transmission such as that decimating sea stars, and also causing birds who feed on different levels of the food chain to die of starvation. You won't get a scientist to make a big statement like that, but it feels like we are watching an egg roll off a counter and nobody will say it's broken till it's on the floor.

A relatively new way of looking at multiple impacts, however, is emerging in something generally called "tipping point" science and explained by Anthony Barnosky and Elizabeth Hadly in their 2015 book, *Tipping Point for Planet Earth: How Close Are We to the Edge?* [1] Hadly is chair of environmental biology at Stanford University and Barnosky is emeritus professor of integrative biology at the University of California, Berkeley. Their career-long research has focused on how changes in the deep past, discernable in the fossil record, can inform what's happening now and into the future. Having studied past extinction events and their aftermath, Barnosky and Hadly warn that anthropogenic pressures on multiple fronts threaten to push ecosystem functioning to a breaking point, a threshold after which what we think of as "normal" may never come again. They list population growth, overconsumption of natural resources, climate disruption, pollution, disease, and "killing anything that gets in our way" as forces that "could push Earth past a planetary boundary that would have devastating consequences. Think, then, what would happen if we exceeded critical thresholds in more than one of them at once." The bad effects each of these impacts is likely to be more than the sum of their individual parts, if "exceeding a threshold in one part of the global ecosystem causes a domino effect of exceeding thresholds in other interacting parts of the system." Which they say is likely from a complex systems perspective.[2]

Back in 2012, Barnosky and Hadly were primary authors on a Nature paper proposing the idea that the threat of planetary-scale tipping points necessitates better monitoring and forecasting of change at both local and global scales. (The paper did not say so, but this is where citizen science monitoring can help and is probably the only tool that can really scale to aggregate big enough numbers of local observations to create a picture of global consequence.) At the time, I spoke with a prominent ecologist who had reviewed the paper for Nature, and while he praised it for the most part, he thought it unlikely that more than one tipping point would be reached at the same time. Recently I asked Barnosky about this criticism. He told me that it has long been held that the various earth systems are "de-coupled" so won't necessarily affect one another. "But things have changed," he told me. "It's us. Humans and human impacts are everywhere—one species has infiltrated every ecosystem. It is less than realistic to not recognize the total connection that is put in place because of this."

Furever

It was still very early in the morning, and monitoring duties would not commence until we reached deeper waters. Jahncke pulled his cap down over his eyes and went to sleep on a bench in the cabin, as did most other passengers. I went and sat with the captain of the *Fulmar*, who navigated by way of a stack of paper maps at his elbow. As we headed farther out, the topography of San Francisco Bay quickly dwarfed the built environment. It was easy enough to imagine what it might have looked like to the first Europeans who came upon the place five hundred or so years ago, a time period that doesn't rate a blip on the geological time scale. A large share of credit for the unfettered view is due to the big protected areas along the coast—together, the Golden Gate National Recreation Area and the Point Reyes National Seashore make up a continuous stretch only lightly dotted with human contrivances like buildings. One even imagines what it might have looked like thirteen thousand or so years ago, the marker contemporary archaeology puts for the date of human habitation in California. The official transition between the Pleistocene and the Holocene epochs is 11,700 years ago—on the one side of which is the last ice age, on the other more or less the climate we know today. But the landscape had different inhabitants then—megafauna, including the saber-toothed tiger and the mastodon. When people came around, they disappeared.

There were perhaps a hundred thousand Native peoples on the continent at the time of the megafauna—how could so few relatively tiny people take down so many enormous and fearsome beasts? Ideas include the possibility that since the megafauna didn't evolve with humans, they didn't know how to avoid predation or to fight back. A related idea is that humans killed enough of the large herbivores to deprive the carnivores of their supper, so the big-toothed starved. Other extinction explanations include climate change—the earth was indeed moving into the temperate phase we have subsequently enjoyed and are in current danger of destroying. And the answer might be: all of the above. Climate-weakened species may have been more susceptible to wipeout by hunting. Hadly and Barnosky posit that the megafauna extinction is an example of a tipping point since "the number of extinctions was multiplied many times over what you'd expect by simply adding up the anticipated effects of a few extinctions by climate change, plus a few more by human impacts."[3] Whatever the cause, the extinction was significant: the Holocene reduction in mammals changed the biodiversity baseline that had existed for millions of years.[4] The megafauna die-off is one of the places historians and scientists place the marker for "end of the world," the threshold past which human impacts have continued to change the geological carbon cycle.

As with the overhunting that may have eliminated all those big animals, so the subsequent whaling and fur trades unconsciously accomplished a biotic cleansing with major consequences to ecosystem health. Today we point to pollution, overconsumption, burning of fossil fuels, and habitat loss to name the forces weakening our biotic support system, but blows have been steadily dealt to the natural world for centuries. The history of North American conquest has fur all over it—that belonging to the terrestrial/aquatic beaver and that belonging to the sea otter. Both critters were hunted nearly to extinction in the service of empire, and their furs fundamentally helped create the sovereign nation of the United States of America. Of course, another story was unfolding at the same time. We were creating and we were destroying.

The Spanish made their way into California in the 1500s but did nothing much about that for the next two hundred years. It was not so easy to send an expedition or to colonize a territory, and the Spanish were busy on other imperial adventures. Sir Francis Drake did make a historic 1579 touchdown here, but the Brits didn't officially visit again for two hundred years. Farther north was left to the Russians, who established a trade in otter skins in the Aleutian Islands in the 1740s. The story of the hunt for the beaver, which systematically wiped out population after population East to West, sea to shining sea, was all about providing skins for the British fad in beaver hats. Otter pelts had a different market: the Chinese mandarin class, avid consumers of otter capes. Otter fur consists of up to a million filaments of hair per square inch, distributed in a smooth, downy layer, and underneath it, fibrous guard hairs. These maintain a membrane of air in the fur that insulates the animal even as it dives underwater, and a perpetual stream of air bubbles accompanies the otter like a bridal train.[5]

Having exhausted the sea otter supply in the North, the Russians started in on the California sea otters, but by this time they had international competition. A period of Wild West mayhem in the waters ensued in the early 1800s, and by the 1830s, American, Spanish, and Russian efforts had completely extirpated the sea otters—they were gone. (In 1841, the Russians sold their West Coast base, Fort Ross, to John Sutter. A discovery near his sawmill on the American River would instigate the next big California natural resource extraction seven years later—the gold rush.)[6]

Invisibly, it would seem, the loss of the otter from the ecosystem began to have an impact. Sea otters are top predators and only *Homo sapiens* hunts them. Most marine mammals thermoregulate, or maintain their body temperature, at least partly through big layers of blubber, but the sea otter (evolved from the river otter) keeps its metabolism steady with that incredibly thick fur, and by eating constantly. A sea otter eats one-quarter to one-third its body weight every day—fifteen to twenty pounds a day for a sixty-pound animal. One benefit of this relative lack of blubber is that sea otters are not tasty to sharks and other big-toothed predators.

Much of the life in the intertidal is about clinging to a surface to withstand the shock from waves—think about your own home being roiled by staggering pounds of energy 24/7. You would hang on to the doorknobs. The abalone preferred by some sea otters stick like glue to rock and human divers need a crowbar or a diving knife to get them off. The sea otter takes a tool more readily at hand, a grapefruit-sized stone, and pounds it until the shell fractures. Dinner is thus presented. The other *spécialité de la* tidal *maison* for sea otters is sea urchins. These are bizarre creatures of ancient lineage and look like fantasy planets doubling as medieval weapons. They are spheres covered in long sharp spines, and at Pillar Point they are embedded in the substrate revealed by super-low tides like mannequin heads at a wig shop. The spines are the urchin's tool for holding on. The spines don't trouble the sea otter, who cracks the hard shell of the urchin and gets at the sushi therein. I asked Pete Raimondi how sea urchins don't get depleted to extinction by sea otters. "They don't get them all," he told me. "The abalone and the urchins that affix themselves deep in the intertidal crevices survive."

The fifteen to twenty pounds of food those disappeared sea otters were now not eating were, of course, eaters themselves. Absent the sea otter, abalone and especially sea urchin populations off the coast of Central California stopped experiencing any *gustatus interruptus*. They got to supersize at the Kelp-to-go, eating ad infinitum, without being controlled by the sea otter. The kelp forest began a devastating decline, analogous to clear-cutting the Amazon or a redwood grove. Giant kelp forests are a feature of the Pacific coast from Alaska to Mexico (and also further south), aided and abetted by that California Current bringing rich stores of nutrients close into shore.

Giant kelp can grow almost two feet a day. For one thing, they don't have to expend any energy withstanding gravity. Cell walls are their only support structure, and their long fronds are evolved for buoyant exposure to the surface of the water so they can feed on sunlight. Kelp have their own attachment issue and manage to stay put by way of haptera, which issue from the kelp in a rootlike way. The haptera maintain a tight grip on a rocky part of the seafloor or else get ahold of a sandy bottom by purposing worm casings into structural elements. New haptera grow on old until the thing—colloquially called a holdfast—comes to resemble a dead tree trunk, providing home to myriad species.[7] The kelp itself provides structure, stability, and food for more than eight hundred species of invertebrates, fish, and other animals. Absent the kelp—well, they all go, with repercussions that extend to the terrestrial realm. Seabirds feed on intertidal offerings, and now their larder is sparser, too.

The good news is that the sea otter came back to the California coast. The story has it that in early spring 1938, a couple who routinely trained a telescope in their home on Bixby Creek in Big Sur got miraculous sight of what they called a "small group of furry floaters . . . rafting in the waves." The outlier sea otters got their numbers back up so that by the 1970s, kelp was waving its green arms in the coastal waters once again.[8]

At what looks to be the halfway point, on Google Maps, anyway, around Monterey Bay between Santa Cruz and Monterey, is a wiggly finger of water penetrating inland and called Elkhorn Slough. Estuaries are bridgers of realms, riverine to sea, freshwater to salt, and the word *slough*, which basically means "muddy," gives you a sense of the terrain. The National Oceanic and Atmospheric Administration and the California Department of Fish and Wildlife jointly operate a research station here, and the public is invited to trek along five well-marked miles of trails through thronging habitat, for, among other creatures, sea otters. Those not inclined to hop on a bike right now to get there can take a look at some of the action going on there by visiting elkhornslough.org/ottercam. You will see for yourself that otters prove that the words *cute* and *top predator* can belong in the same sentence.

Noodling along in the otter world, I had heard from several unofficial sources that sea otters used to spend time on land, but having learned the hard way that humans liked to make them into capes had decided sometime back in their genetic history that it was better just to stay in the water. This conventional wisdom was in fact breached a few years ago when two volunteers at the Monterey Bay National Marine Sanctuary at Elkhorn Slough decided to go kayaking together at 2:00 AM—to spy on sea otters. Ron Eby, a retired naval commander, and Robert Scoles documented healthy animals "hauling out," or resting on beaches, something it was assumed only sick or injured sea otters did. At first nobody listened to them, until Daniela Maldini, then director of research for Earthwatch Institute, helped them collect their data according to science-driven protocols.[9] "It gave us credibility," said Eby, who has logged more than ten thousand volunteer hours at the slough. Working together, volunteers and scientists have helped establish that yes, the slough is valuable habitat worthy of protection.

One more thing about Elkhorn Slough. In the late nineties the Elkhorn Slough Foundation, raising money to support research and restoration, decided to become a land trust. They were determined to "acquire, protect, and manage" the resource. Ultimately this is where citizen science has to go. Just producing data for scientists to write papers with, and even the magnificent emotional and intellectual returns of direct observation and care of nature, is not going to save it. Saving nature means protecting habitat. Most of the time that's going to mean finding ways to buy it so it doesn't get developed. Go, Elkhorn Slough!

Kelp may be enjoying the return of the sea otter, but it still has big problems. For example, kelp forests have been diminished by more than 80 percent in the past century in Southern California. Weather events like El Niños rip it out, and that's natural enough, but pollution, development that causes sediment to flow into the water, overfishing, and still-paltry numbers of sea otters compared to their historical numbers are to blame. According to the California Coastkeeper Alliance (CCKA), a single discharge of heated water from the San Onofre Nuclear Generating Station in San Diego County destroyed more than 150 acres of kelp forest in one fell swoop.[10] And now with the sea star wasting die-off, proliferating sea urchins have been relieved of another top predator and are busy decimating their own food supply.

Citizen scientists have been on the case with a giant kelp restoration project led by the CCKA. The usual arsenal of inventorying and monitoring are well in hand on this restoration project, which has engaged more than one million community members in Southern California thus far. Citizen scientists manually remove sea urchins from kelp beds, plant lab-grown kelp, and transplant drifting kelp, which turns out to be easier and more successful than the lab-grown alternative.

Eyes on the Skies
Most citizen science projects having to do with biodiversity want you to get some contact with the great outside and do something with it, even if it's only on your windowsill: watch for pollinators at prescribed times of day, for example, or, as with the kelp rescuers referenced above, wade into the water

and help garden the ocean. A significant platform in the citizen science world works otherwise, however, and participants can volunteer on many projects without leaving the house. With all those people glued to their computers in enclosed spaces, it is no wonder the project is called Zooniverse.

Zooniverse was born with the 2007 Galaxy Zoo, a project now in its fourth incarnation, and the idea is to mob-source not data collection but data analysis. Participants logging into Galaxy Zoo on their computers are asked to discern patterns in galaxy images—photographs from space—and mark down what kind of shapes they see in them. In the case of finding new galaxies in the universe, participants looking at pictures are directed to find shapes—the human eye is better at discerning the elliptical shape of a galaxy than computers are. But Zooniverse provides the data to the user—imagery of the cosmos from the Hubble Telescope. "You can think of Galaxy Zoo as a cosmological census, the largest ever undertaken, a census that has so far produced more than 150 million galaxy classifications," is how "open science" advocate and an inventor of quantum computing Michael Nielsen puts it in his book *Reinventing Discovery*.[11]

In May 2012 the then director of Zooniverse, Arfon Smith, addressed the crowd at a citizen science convening held at the California Academy of Sciences. Pretty much all the projects represented from around the country had to do with monitoring and inventorying: with counting birds, identifying invasive species, measuring this, or documenting that. "I'm glad we're in our shoes and not yours," Smith told the group. "How hard it will be to build a robust data set from nonexperts. Data analysis is something we understand. We build the dumbest thing that works, and that's what we try to distill a problem to."

"The minute computers are better at a task, we stop asking people to do it," Smith said. "We didn't set these projects up to teach people about our subject; there's a much better article on Wikipedia about universe morphology than the one we provide." Humans are naturally bent toward identifying anomalies, in what looks different than expected, even while engaged in a mundane sorting exercise. In Smith's description, Zooniverse is like a depot for excess brain-cell functioning. "Adults spent two hundred billion hours a year watching television," he pointed out. "There is a surplus of brain time to solve problems."

Smith's talk at the academy both energized and slightly depressed the nature nuts in the room. All those lugabouts out there, who could be saving nature!

Elizabeth Babcock, chief public engagement officer and dean of education at the academy, said, "Okay, you have people observe, classify, transcribe. You ask them, 'Is it there or is it not?'"

Smith replied, "Yes, it's a measuring thing."

Babcock, ever sharp as a tack, retorted, "Well it's a transcription thing—that's copying." Not only the galaxy work but most Zooniverse projects are built on this program.

Getting the actual application of her kind of citizen science, Babcock remarked, "We have another purpose—conservation."

Smith pointed out that as data sets get more massive and photographs, used as data points, become more ubiquitous, we're going to need to deploy those excess brain cells to make some sense of them.

Couch potato citizen science has its pleasures. My son and I periodically sit in the Sky Oaks Ranger Station on Mount Tamalpais in Marin County and stare at computer screens, discerning species that have triggered motion-activated wildlife cameras and making note of them. Three hundred cameras have been arrayed on Mount Tam according to protocols established by the Zoological Society of London and the Wildlife Conservation

Society and called the Wildlife Picture Index. So we sit there and stare at frame after frame, making note of who is coming across the lens when (each photo has a date and a time associated with it). One day I looked at hundreds of photographs that chronicled the days and nights of a wood rat. In the morning the little mammal would seem to come out to go grocery shopping, making a quick semicircular arc around a tree. In the evenings perhaps the same animal or maybe its mate frequently seemed to make a dash for something—maybe getting home after a long day's work out in the underbrush. That day my son was getting all sorts of sightings—an owl, a deer, a bobcat—but I was mesmerized by my one spot on the mountain, feeling I knew its denizens in some ways like I know those in *The Wind in the Willows*.

To those of us inclined to feel that the stars are taking care of themselves while the tide pools are under assault, Floating Forests comes as a welcome meeting point of earthly and celestial. This is one of Zooniverse's most recent projects, in which the participant helps identify not cosmic gases but kelp. The project began with researchers studying the connection between waves, kelp, and food webs, but their study subject kept getting washed away by big winter waves. Since divers went out only in the summer, they couldn't estimate winter storm damage to the seaweed. The solution was to use satellite imagery. Since 1972, NASA has captured spectral information from Earth's surface from space, at a fine enough resolution that land-use changes can be recognized. As the satellites orbit, every area of the earth gets photographed about twice a month. Kelp is visible in the pictures but computers can't tell it apart from land or sea foam. Human eyes, however, can discern patterns in kelp by where it is and how it clumps. The researchers convened a "great pipeline" by which they fed images to undergrads who circled kelp on the images and forwarded the information to grad students and computers for further analysis. But, as Floating Forests' Jarrett Byrnes writes on the project's blog, "It took years just to get California." The idea of analyzing the whole earth, and using the historical Landsat databases that go back to the seventies, "seemed impossible." Until Zooniverse stepped in and helped them reach thousands of citizen scientists who are helping them classify the kelp. (You can help: floatingforests.org.)

Seen and Unseen

Since sea otters are a top predator in their aqueous realm, the unraveling of the ecosystem upon their depletion has been fairly apparent. The vast reduction of whale populations is having impacts that are less evident to initial observation. Science, including paleontology, which looks at the very long time scale, is elucidating the consequences of their enormous absences.

The blue whale is the heaviest of any animal that ever existed, and the species is still around. Even as whale populations hang on by threads here and tatters there, they do still swim in the deep. Our oceans today have plenty of problems, but it is harder to drive animals to extinction in the water than it is on land. While we were pretty well able to round up and exterminate nearly every last bison in North America (populations have been cultivated and supported in some areas), we could not quite do the same thing to the whales—they can swim away. The commercial whaling business for a millennium operated in the dispiritingly consistent way we have reduced and eliminated other commercially valuable creatures, depleting first coastal populations and then pelagic, or deep sea, populations of whales.

The whaling era epitomized by Herman Melville's *Moby-Dick* is deeply associated for many of us with the East Coast, and Nantucket, New Bedford, and Sag Harbor were historically centers of the trade. The East Hampton house I grew up in has a "widow's walk" on the top of it, a white-post lookout from which those at home could anticipate the return of their whaler kin. Whaling took a distinct turn to the left coast beginning in the late 1700s, when the first nonindigenous Pacific whalers rounded Cape Horn and headed up the coast of South America. They returned to New England with cargo loads of sperm whales from the coast of Chile. The evident abundance of whales instigated more Pacific ventures, which crept ever northward, reaching the equator by the War of 1812. After a three-year martial hiatus in the pursuit of whales, the industry resumed in 1815. From 1830 the Nantucket fleet of whalers went increasingly to the Pacific and, in 1838, came west.[12] The first whaling vessel outfitted not from the East but from San Francisco was sent to the Galápagos in late 1850, and in 1851, shore whaling was instigated in Monterey. For two decades shore whaling went on from eleven stations along the coast of California.

In 1858 Charles Melville Scammon (not related to Herman Melville) put his bloody signature on the history of Pacific whaling when he sailed his brig *Boston* into a lagoon off the Sebastián Vizcaíno Bay on the Baja Peninsula. When the brig happened upon hundreds of female gray whales and their newborn calves, two of the adults were dispatched. The evidently easy pickings were about to turn the tables on the whalers, however, and when the boats were lowered again the men aboard were . . . whaled. The whales fought back, smashing two boats. A few days later Scammon deployed his men again but most of them jumped into the water as they got within range of the whales. For a brief moment, Scammon gave up whaling with what he called his "panic-stricken" men, but then he came up with another idea. Putting to use a bomb lance with a delayed explosion, he resumed dispatching the whales. Competitors joined the fray and soon "Scammon's Lagoon" was depopulated. And soon after that, the California gray whale was nearly completely extirpated. In 1874 Scammon published *The Marine Mammals of the North-Western Coast of North America,* in which he waxed poetical: "The mammoth bones of the California Gray lie bleaching on the shores of those silvery waters, and are scattered along the broken coasts." And he wondered "whether this mammal will not be numbered among the extinct species of the Pacific."[13] In large part thanks to him.

The gray and bowhead whales were at the brink of extinction by the mid-1800s. Other species of whale would be pushed to the end of their lineages later, by the so-called industrial whaling that started up after World War II. Human conflict provided a hiatus for the giants of the deep—people were too involved with killing other people to go after them. But when whaling resumed it did so with a vengeance. High-speed "catcher" boats and mechanized factory ships hunted down their prey with horrible efficiency. Where eighteenth- and nineteenth-century whaling had been in pursuit of oil to burn as an energy source, with a significant market for whale bones with which to make corsets, twentieth-century whaling was about hunting for whale meat for eating. In the 1970s, my aunt Marie decorated her every Christmas and birthday card with a couple of happy little cetaceans, and in her perfect cursive cheerfully exhorted: "Save the whales!" The quick devastation visited on whales by industrial fishing was met with national and international sanctions on hunting them that have allowed some populations slowly to come back. Hunting of blue and humpback whales was halted in 1965, sei and fin whales in 1975, and sperm and Bryde's whales in 1979. In 2014 blue whales in California waters had rebounded to roughly historical population levels. Humpback whales have also had a comeback. Blue whales elsewhere have not done so well, and the situation is still grim for most whales worldwide.[14]

Whale numbers have been almost incomprehensibly reduced in a short period of time, and science has just in the past few decades been able to quantify what losing them has done to the ecosystem. Losses of big players help drive further negative consequences for the life that's left behind. Contemporary estimates put the number of whales killed in the thousand years of commercial pursuit of them at tens of millions. To look at the problem another way, whale biomass has been reduced by 85 percent over time.[15] The horror of the focused devastation of a magnificent historic animal notwithstanding, this removal of the ocean ecosystem's largest-bodied creatures cannot fail to have had an impact on the way things work in the water.

Whales are not just floating around, spouting, and whacking whalers with their fins. They are players in the food web. They eat things; they are eaten by things. (Not too many animals aside from humans are capable of predating on whales, but killer whales do feed on some species of whale.) Whales are big, moving food trucks, embodying energy and nutrients and distributing these in deep waters that lack them. Their significant poops fertilize the water. As Craig Smith, professor in the Department of Oceanography at the University of Hawaii, told me, "They essentially sustain their own food supply by providing a nutrient-rich environment for krill to grow in." Plenty of other species also avail themselves—and depend on—krill, which are essentially at the base of the ocean food chain.

Smith is an expert on the poignantly called specialty "whale falls." When they die of causes other than being hunted by *Homo sapiens*, whales usually, yes, fall to the bottom of the ocean. Some of course are beached, in which case, in the parlance of biologists, they give a nutritional boost to the terrestrial food web—when scavengers are not stopped by well-meaning people from eating the carcass. Can you just picture a gigantic leviathan falling to the deep, deep bottom of the sea? It's actually bringing massive cargo-loads of nourishment to a realm in which, lacking the ingredients of organic interaction, the cupboards are ordinarily bare.[16] Whale falls are analogous to tree falls in the forest: the dead one becomes a habitat for many more species. It's moving the materials of life around so that more can partake of it; whales support biodiversity both while they are alive and when they naturally die. Whale falls from the North Pacific fossil record go back thirty million years—which gives us a bit of perspective on how long these leviathans have been fertilizing the ocean and helping sustain other life-forms there. The removal of whales out of the ocean for human purposes has deprived up to 350 species that have coevolved with whales, dependent on their carcasses, of hearth and home. It is postulated that the first marine invertebrate extinctions resulting from human impacts are traceable to the plucking of all those large bodies out of the sea.

In the press on climate change one often reads and hears about plans to "sequester" carbon, or get some of the excessive amount of CO_2 we are putting into the atmosphere incorporated, say, in the trunks of trees. All life-forms are made of carbon, so whales, weighing in at 30 to 160 tons, are nontrivial stores of carbon. Whale falls even now transfer about 190,000 tons of carbon a year from the atmosphere to deep waters where they eventually decompose and become part of the geological carbon cycle again. Restoring whale populations to historical numbers would equal sucking down carbon in amounts equivalent to climate-engineering schemes, and of course would do so naturally, with many further ecosystem benefits we can surmise and probably many more we don't know about yet.

There are other ways to look at the loss of big-bodied animals over time, which have to do not with sequestering carbon but with releasing it into the atmosphere. That 85 percent of whale biomass eliminated by hunting not only did not get to decompose in the ocean and so nourish other life-forms; those whales were also burned up as oil or consumed as food by us. Fuel burning of course releases carbon into the atmosphere. Compared to the amount of coal, oil, and gas with which we are currently fueling our industrial world, the vaporized whales don't add up to so much—they are puffs of smoke. But their loss is more than that. They could be sequestering carbon and enriching other life-forms.

Although both the whale and sea otter trades were fundamental to establishing European settlement in California, and indeed the entire continental expanse, there is a critical difference between the two resources. Sea otter pelts kept people warm the old-fashioned way—they wore them. Whale oil contributed one of the first energy sources by which we powered industrial expansion. Today, of course, we rely on fossil fuels to do the trick, and that's arguably our biggest problem at the moment. In 1896, the first US offshore drill was built just south of Santa Barbara to access stored energy. Fossil fuels, to remind you, were once animals and plants whose bodies were made possible by photosynthesis. Decaying as the whale falls do, over millennia they became transformed into coal, petroleum, and natural gas. Once we figured out how to purpose these stores of energy, the human race was off to . . . the extinction races.

I See You

Having reached a predetermined latitude and longitude, the captain of the *Fulmar* idled the boat and, rousing from their naps, the crew began to get to work. We were mostly alone, it seemed, on the ocean, but an occasional fishing boat came into our view and on the horizon, slow-advancing cargo ships the size of city blocks. The first marine mammal we saw made its presence known by leaping with wild and sustained enthusiasm at the prow of the boat. It was a sea lion, unlike those I was used to observing from shore, which are much larger and pretty much stay submerged except for an occasional heads-up. This little one jumped and twisted as if for joy and to say "Hi! Hi! Hi!"

As we got farther out into the ocean there was suddenly another herald, this time in the form of giant gushes of sea spray. It was not one humpback whale; it was two. Jahncke pulled out his smartphone and made the Whale Alert observation. "So hopefully these two whales won't end up in the propellers," someone remarked.

As aforementioned, in addition to doing the usual counting-up of birds and mammals and sampling of the watery depths, Jahncke was also test-driving the smartphone application Spotter Pro, a citizen science app that will subsequently be used to transmit any whale sightings to sanctuary managers, US Coast Guard officials, charter fishing boat operators, whale watchers, and other shipping captains. In the words of John Berge, spokesperson for the Pacific Merchant Shipping Association, "No ship captain wants to hit a whale."[17] But too frequently, they do. Having been hunted to the brink of extinction by the 1970s, since then most whale species have been protected and many of their numbers are slowly improving. However, a worldwide problem has emerged, and that is the collision of whales and especially cargo ships plowing their way into and out of commercial ports. The Port of Oakland, in San Francisco Bay, is one of the country's busiest. Between 1988 and 2012, at least twenty whales were confirmed killed or injured by ships near San Francisco, and this number is estimated at less than 10 percent of the likely total of ship strikes.[18]

The immediate purpose of a Spotter app observation is to help alert ships to whales present in the mile-wide shipping lanes that organize traffic into and out of the bay, so that they can either switch lanes or slow way down until the whales are clear. The app feeds data into a database called Whale Alert, now in wide use by whale watchers and lots of other people out in the water. Most citizen science projects burgeoning today have some group or community element—the three-hundred-plus members of the Golden Gate Raptor Observatory, of which I am one, for example, meet regularly and have an annual awards dinner. Whale Alert doesn't have any intrinsic identity beyond this one piece of utility—see a whale, make an observation—but it takes advantage of the eyes out on the water to make a better aggregate picture of when and where the whales are.

The first citizen observation using Spotter Pro on the West Coast came within twenty-four hours of its release. Jahncke emailed me: "The citizen scientist reported an observation of a humpback whale within the precautionary area where all shipping lanes converge prior to entering the San Francisco Bay." Within a week, and despite thick fog, sightings were reported of five humpback whales and fifteen harbor porpoises in the Greater Farallones National Marine Sanctuary; in Monterey Bay, nine humpbacks and three porpoises; and in the Channel Islands, two more humpbacks. By 2015, what was updated to become Whale Alert 2.0 had been deployed by nearly four hundred contributors, and more than 1,350 whale sightings helped the big ones stay propeller-free.

As a blue whale was sighted I clung to the side of the boat waiting to see it arise. What looked to be an underwater plain uplifting, then receding back under the dark waters, sustained a liminal presence next to us off and on for about a half hour. Blue whales tend to swim in the depths and it's more unusual to see them breach than it is other species. Enough of its vast breadth was visible to notice how smooth the blue whale is, compared to the lumpy, barnacle-coated humpback. The way it came to and retreated from the surface was like a dream. Each time I discerned its oblique presence, making out just a fraction of its submerged enormity, was like a peripheral miracle. You are there, blue whale, *you are there.*

Eyes Wide Shutdown

Chatting between counting and monitoring duties aboard the *Fulmar*, the conversation dipped around again and again with dread to the looming shutdown of the federal government, which was fated to occur between October 1 and 16, 2013. Jan Roletto, who founded the Beach Watch program in 1994 in its current form in response to the 1989 *Exxon Valdez* oil spill, was particularly anxious.[19] Roletto helps manage more than 120 citizen scientists who comb forty-one beach segments in the Bay Area every two to four weeks. Although they don't get paid, Beach Watch volunteers are covered by workers' compensation insurance provided by the federal government. Because this would be suspended in the event of a shutdown, Roletto would be enjoined to tell her volunteers not to go out on their appointed rounds. Twenty years of consistent data on the state and health of the beaches—reflected in the status of seabirds and marine mammals—would have a lacuna in it. This could have a negative impact on future litigation and settlements regarding oil spills. The whole reason Beach Watch was founded was to have incontrovertible data points collected and archived according to strict protocols that hold up in court.

Oil spills have played a big role in conservation over the years. When Platform A in the Santa Barbara Channel experienced a blowout in 1969, the horrors of 3.4 million gallons of oil spreading over thirty-five miles of shoreline prompted the establishment of the US Environmental Protection Agency. Oil spills from tankers moving in and out of San Francisco Bay have had particularly devastating consequences as far as wildlife and the health of the water are concerned. In 1984 the tanker *Puerto Rican* leaked 1.5 million gallons of oil outside the Golden Gate, killing five thousand birds.[20] In 1988, a Shell Oil storage tank spilled 420,000 gallons into wetlands near Martinez. After the infamous *Valdez* spill, when lawyers for Exxon successfully argued that the company should pay relatively low damages in response to bird mortality because no one could prove how many birds were around before it happened, citizen-powered monitoring groups all up and down the coast were marshaled to start counting in preparation for the next spill. Beach Watch volunteers count birds and mammals, and photograph any dead animals they come across. They also collect any pieces of oil they find. Analysis can trace every bit of oil to its source.

Beach Watch volunteers undergo eighty hours of training before they can begin monitoring, and the program requires periodic retraining. Volunteers will tell you the rigorous training is a main motivation for participation. Every piece of data is taken into "custody," established by written forms that express who has responsibility for it when. This is a routine for maintaining the integrity of evidence to be used in a legal proceeding. Because Beach Watch and other surveys like it follow this letter of the law, and because every observation is verified by a photograph that is then verified by an expert, there is not as much discussion about the reliability of this data as there is regarding that collected by other citizen science programs.

In 2014, a group of about fifteen Beach Watch volunteers celebrated their twentieth anniversary combing Bay Area sands, documenting birds dead and alive, and hunting for evidence of oil spills. Going out with many of them on their appointed rounds, I found this otherwise heterogeneous bunch uniformly soulful and serious. Resolutely place-based, they were intimately attached to their beaches, knowing every square inch of sand it seemed, and regaling me with details of erosion and water-level patterns over time. Photographing the picked-over carcass of a Western gull, one volunteer pointed out to me its resemblance to an Anselm Kiefer painting and discussed the iconography of holocaust. "A large part of what we're doing out here is noticing death," he told me. "Does it get you down?" I asked him. He handed me his binoculars, and directed my gaze to a squadron of pelicans flying over the horizon. "They're back from the brink," he said, referencing their once perilously low numbers. "We're out here counting but we're really fighting for all of this," and his arm swept the beach.

To the north, Dr. Julia Parrish's Coastal Observation and Seabird Survey Team has been going since 1991. Parrish has between 750 and 800 volunteers monitoring beaches from Northern California to Alaska. "When we talk about science," she told me, "a lot of people talk about the scientific method. This is a question or a wondering you turn into something you can test, a hypothesis. But before we get to the question and the test, we go out and do some research to find some pattern, a taxonomy, a relationship, and once we establish that pattern we ask, 'How did this come about?' A lot of science is going out and establishing a pattern."

When the White House conferred on her the honorific Champion of Change, Parrish commented, "Scientists alone can't begin to document what's normal, let alone how fast things are changing. We need a willing army to make that happen. In short, we need citizens—the locals who watch, and know, and love their backyards, their environments."

"When people tell me, 'I want to do this,'" Parrish told me, "I think, no you don't. The threat of avian influenza or a big rack of dead birds will get a lot of media coverage and people will say, 'I want to help.' And you know they've watched a lot of nature shows and think it's going to be a lot of sex and violence everywhere you look and they'll be in the middle of it. And when they actually go out, it's boring. I'm looking for the person who is fascinated by the normal. I want the person who walks out not with histrionic anticipation but with a quiet sense of 'I wonder what I'm going to see today.' We have data on this. People go out to places they love. They build up a model of expected changes, as with the season. They are not novices. They are experts. These are the people I want. They are invaluable."

You Are Invading My (Evolutionary) Space

As I watched from the deck of the *Fulmar*, the Farallon Islands themselves hove into view. Named by Juan Rodríguez Cabrillo in 1539, they are, as the Spanish word *farallón* indicates, steep rocks jutting up from the water so stark they look like cutouts. As I conducted interviews for this book, several researchers asked me if I planned to visit them, something you have to get permission to do through the US Fish and Wildlife Service, which manages the islands in partnership with Point Blue. I haven't managed to get USFWS to say yes I can, though they haven't come out and said no either. I guess in an excess of honesty, I have been very clear in my appeals to USFWS about why I want to visit the islands, which has to do with mice. Making a fuel stop at the Farallones, migrating burrowing owls find the densest population of house mice recorded anywhere in the world. That's a lot of dinner easily served, so many of them decide to stay put. In plague-like numbers, the mice are said to make a moving carpet when the population is at its peak. The population goes through a natural crash in late fall or early winter, at which point the owls turn to the ashy storm petrel, a scarce, tiny seabird. The bird is being hunted by the owls to near extinction. Worldwide, there are an estimated eight to twenty thousand ashy storm petrels, and about half of the breeding population do so on the Farallones. It is widely believed that unless the mouse is eradicated and the owls thus dissuaded from dining there, the petrels will shortly become locally extinct. This would deal a significant blow to the vitality of the whole species.

And why do I want to report on this? Because this is a classic case illustrating invasion biology. Invasive species are probably the number one immediate threat to biodiversity today, and although this is not a case where citizen scientists can help, there are many other invasive situations that depend on the early-warning systems of volunteer observation, and also on volunteers yanking invasive weeds or otherwise plucking invasives out of systems where they don't belong. The invasive mouse on the Farallones is most likely a remnant insult brought inadvertently by the generations of hunters who harvested the islands' bounty to sell on the mainland. When another invasive population descended on San Francisco Bay in pursuit of the gold rush, collectors known as "eggers" harvested millions of common murre eggs to feed the newly enlarged population of *Homo sapiens* over a period of about forty years, nearly eliminating the bird before their practices were halted.[21]

Point Blue scientists and many others advocating for eradication of the house mouse are considering the use of a rodenticide to do the job, and this is unusual for people who are passionate about birds. Even Allen Fish, head of the Golden Gate Raptor Observatory and a bird booster if ever there was one (more on him later), supports the plan to eradicate the mouse on the Farallones. Typically birds suffer collateral damage from rodenticides. The pros, the cons, and the possibilities of how to do this have been and are being agonized over. "The level of site-specific scientific analysis, and consideration of impacts on 'non-targets' and mitigation impacts to minimize those, are at a level pretty much unseen in the history of these projects around the world," Russell Bradley, senior scientist and Farallon program manager for Point Blue, told me. Bradley explained that a novel plan to "haze" or basically shoo away the proliferant seabirds would help prevent them from ingesting the poison themselves while it percolates through the food web—it's currently under consideration by USFWS. Successful nonlethal methods to shoo gulls away include playing raptor attack calls (biosonics), passing a laser briefly over a colony (they all take off), and hanging dead gull effigies from poles ("Sounds weird," Bradley said, "but it's highly effective."). But let's face it, USFWS is probably not too keen on my enthusiasm for giving the issue more publicity. The idea of dropping rodenticide on a national marine sanctuary is too much for some to bear.

Continuing the Journey

The dynamics of invasions and what they do to ecosystems is understood by way of conservation biology, a science officially established a mere thirty or so years ago to do as the Latin source word indicates: *conservare*, to save, nature. Conservation biology, in turn, has developed from the observations and ideas of Darwin, Wallace, and their precursors. The terrain is the pattern and process of nature, and key conservation biology concepts include island biogeography, connectivity (the idea that swaths of nature need flow of species between them), trophic cascades, species ranges, and coevolution. These concepts are interconnected and inextricable from one another. The discovery process behind them—the stories about how scientists have come to figure these things out—represents the sort of science that is generally not undertaken by those without a PhD or not working to get one. The science of how life-forms have come into being and dispersed around the globe provides a blueprint for how and why we need to do citizen science. And citizen science focused on invasions is expressly at the intersection of evolution and ecology. One of the most effective and well-run citizen science projects in the US, Vital Signs (vitalsignsme.org) connects researchers who are grappling with the impacts of invasives on land and water with citizens (often students) who essentially do their field work for them. The citizens are also making up an early warning system for natural resource managers who can often eradicate invasives before they completely take over.

Like the Galápagos tortoises, every species has a creation story—how it came to be where we find it. Jami Belt, who leads several species counts in Glacier National Park, told me, "Sometimes a volunteer will work with us for months and then put up a hand and say, 'Why are we counting loons, anyway? Don't you scientists know how many birds there are?'" Citizen science project leaders don't often have time to explain that species are distributed in different amounts around the earth, and keeping track of how many loons, for example, arrive at Glacier each summer is a way to keep a finger on the pulse of how the species' populations are faring, and why, and how what the loons are doing today relates to their evolutionary history. If the populations are going down, down, down over several years, are the birds declining in number or are they seeking a different habitat? Where are they going, and what is that telling us about broader changes in the ecosystem?

For millions upon millions of years, species have moved around the globe, setting up shop, and when they have become isolated from others of their own kind, they have evolved into new forms. The parallel narrative is that when a threshold is reached of too much isolation from others of their own kind, a population goes extinct. This takes a bit of explaining, starting with Darwin (and Wallace), and hopscotching to E. O. Wilson and his colleague, mathematical ecologist Robert MacArthur.

Darwin and Wallace were not the only ones pondering the mechanisms by which new life-forms come into being, and science got even busier on the subject when *On the Origin of Species* was published in 1849. But it took another hundred years before science grappled with the real-world application of the concepts of natural selection, speciation, and extinction on the landscapes you and I tread over. As much as Darwin in particular, with his reams of observational data, replaced the story of biblical creation with a narrative based on actual nature, by today's standards his work remains in the realm of lore, or storytelling. The discipline that eventually evolved to quantify the story and makes it testable is conservation biology. In the telling of many, a single book kick-started the field: *The Theory of Island Biogeography* by E. O. Wilson and the late Robert MacArthur.

Ant Man

Since he's arguably the most famous of living scientists, not only revered for seminal contributions to evolutionary biology but also beloved for his generosity of spirit, it's kind of hard to imagine E. O. Wilson feeling envious and insecure. But the time was the late 1950s and early '60s. Wilson was already established at Harvard, regarded as a prodigy in the field of natural history. And that was just the problem. Knowledge of biology was undergoing a revolution, and Wilson's branch of the sciences was resolutely out of the loop. A new way of looking deep into organisms, via microscopes and known as molecular biology, was all the rage. Wilson's specialty, entomology, was rendered quaint, and so were its sister disciplines, zoology, ornithology, botany, and so forth. As Wilson put it in his memoir, *Naturalist*, "The science was now being sliced crosswise." And he admitted that the level of the molecule is a pretty nifty place to look at life. All science seeks what are called "general principles," or simple ways to explain things. Gravity, for example: what goes up must come down. Evolution, for another: the fittest live. All organisms are made up of molecules; ergo, here is where unity and diversity can best be understood.

As he was being elbowed by "test-tube jockeys" proliferating like the *E. coli* that was their favorite study subject, the man who would otherwise prefer to crawl among the lowliest of bugs had to muster some major moxie—even arrogance, if one considers that the declaration of a new theory in science is tantamount to declaring a new religion and oneself its god. But Wilson was not only in need of saving his own career; he felt charged to rescue a whole field. He knew that even molecular biology occurs within the broader context of place and time, or biogeography, and he saw that this initial rash of molecular biologists, so eager to chuck the organism and its address in favor of its tiniest parts, were throwing the baby away with the bathwater. How could such an important and central framework seem even for a moment dispensable? Some soul-searching was in order.

One problem Wilson confronted was that evolutionary biologists hadn't been connecting the dots between their individual study areas and weren't contributing any broad synthetic thinking to their own field. Evolutionary biologists "could point to no recent great advances comparable to those in molecular and cellular biology," and hadn't brought comparable rigor to their own investigations.

Hatching his plan to redress the situation, Wilson connected with MacArthur. In the words of historian Donald Worster,

MacArthur had "'superbrain' charisma," and lived by the mantra "Predict, predict, predict." "An ecology that makes no testable predictions," Worster notes, "is not yet a science."[22] If evolutionary biology was to be taken seriously it would have to quantify its findings, to render its postulates verifiable, its experiments repeatable. If it was to be considered real science, lore, or "mere metaphors and worldviews," must be eliminated, replaced with more math. The way to do this partly entailed "keeping a precise body count" of species, or data points.

Taking a significant page from Darwin and Wallace, MacArthur and Wilson focused on islands as a way in to understanding biogeography. Islands are by definition separate areas, and it takes some time, but you can actually count up the members of each population of many of the species on an island, and you can track these over time. You can also measure the area of an island, and then you can compare the number of species, the size of each population, and the size of the island. The "species-area curve," as MacArthur and Wilson hammered it out, is still in use today as a predictive model of how many species you will find on an island. In broad surmise, the bigger the island, the more species, until "equilibrium" is reached, at which point the rates of extinction and immigration reach parity. The smaller the island, the more extinction you will have—if it gets slightly warmer, or colder, or a newcomer arrives, there will be fewer places for species to persist. The closer an island is to a mainland source of species, the more immigration you will have, and the converse. MacArthur and Wilson proposed that overlapping the immigration and extinction rates allows you to calculate the number of species on an island.

In a brief aside in their very brief book, MacArthur and Wilson noted that as human impacts are creating "islands" out of natural areas on continents, their theory could be useful for conservation. They illustrated this by way of a series of four maps made by John T. Curtis, then a botany professor at the University of Wisconsin. What begins in 1831 as an almost completely dark shadow, representing forest, becomes progressively broken up by little white bits indicating human development. By 1950 Cadiz Township (Wisconsin) looks like an x-ray image of the original—only little dots of naturalness remain in a sea of white, and those remnant pieces of forest look exactly like islands. Essentially, human impacts make nature into smaller and smaller islands farther and farther away from sources of renewal for their populations. The highest rates of extinction occur on islands, and the biggest driver of those extinctions is invasive species.

It's All About You

A graduate student working with Wilson back in the day, today Daniel Simberloff is best known as the high priest of invasion ecology, which he came to as "an intellectual exercise. It occurred to me that introduced species ought to be a probe into the various theories of community structure, like the equilibrium theory of island biogeography, which says that if one species arrives, another goes extinct." There is a range of definitions for *invasive species*, but generally these are regarded as newcomers that did not evolve alongside the other species in an ecosystem but were brought there mostly through human activity. Invasives spread rapidly, often outcompeting native plants and animals and eliminating them directly. They can also cause further levels of extinction as time goes on.

Through the 1970s, Simberloff tested the strength of island biogeography theory, finding that the introduction of invasives did not necessarily lead to a quid pro quo with other species, unless the invasive was a predator, in which case extinctions often followed. But in the mid-1980s, his approach to the matter of invasives shifted from the theoretical to the actual. "I was asked onto the national board of governance of the Nature Conservancy. I'd go to their meetings four times a year, and we would hear from land stewards," people who either owned or managed significant tracts of landscape, "and they would present to the board. I couldn't help but notice that surely two-thirds of the problems they reported were introduced species affecting the functioning of their holdings." The land stewards not only reported their woes, they sought Simberloff out for advice. "I thought, 'Wow, this phenomenon is not only of academic interest—it's a real-world problem.'"

Today, of the big global change agents that are accelerating species extinctions, biological invasions are perhaps the most in-your-face. All over the world, communities struggle with the impacts of invasives that eliminate natives directly, including introduced mammals that wipe out smaller animals, herbivorous insects that decimate agriculture, plant pathogens like the chestnut blight that wiped out entire forests in eastern North America. Invasives can also cause further levels of extinction as time goes on. They are an emphatically practical problem, like the steady takeover of the Sonoran Desert by buffelgrass, which threatens to choke out the iconic saguaro cactus in the American Southwest. In response to the real estate, recreational, ecosystem, and other consequences of invasives, billions of dollars are spent every year to combat them. Invasions don't only affect what you see aboveground, but reduce native biodiversity from bacteria to fungi, and even eliminate half of the nutrients in the soil over which they rage.[23]

Botanist Joseph Hooker, Darwin's close friend, noted on an 1890 trip to the island of Saint Helena off the coast of Africa that goats and introduced plants had come to dominate the island's native ecosystem. Darwin himself wondered whether the most successful invasives might be those that do not have any relatives in the habitat in question; now called Darwin's naturalization hypothesis, his idea is being proven by molecular investigations of species relatedness. The late English zoologist and animal ecologist Charles Elton based his 1958 book *The Ecology of Invasions by Animals and Plants* on a series of BBC broadcasts he gave called *Balance and Barrier*. Elton set the historical stage for understanding invasions by explaining how plants and animals evolved in relative isolation on "Wallace's realms," or those six distinct biogeographic regions of Earth defined by Wallace. Elton's book is easy to read and often charming, and his basic insight still holds true: "The real thing is that we are living in a period of the world's history when the mingling of thousands of kinds of organisms from different parts of the world is setting up terrific dislocations in nature." Most invasions can be traced to the peripatetic journeying of *Homo sapiens*, which we have been doing for thousands of years. Globalization has sped the plow in recent decades and brought us to our current state of nearly ubiquitous invasions.

Invasive species narratives often play out like horror stories, leading in some cases to what researchers call "ecological meltdown." While some invasives are notoriously difficult to control, the quest is not impossible. Citizen science projects all over the world have been set up to document and monitor invasive species, and even those like iNaturalist (about which more later), not expressly invented as early warning systems, still do function as such. Often a total invasion by a noxious newcomer can be averted if the first few instances are eradicated quickly. Different kinds of invasions require different strategies; for example, eradication of rats from an island is an all-or-nothing affair, and every single last rat must be eliminated if the action is to be effective.

The Call to Action

One of the world's most successful nature-protection organizations today is Island Conservation, and its creation story, or heroic journey, goes like this: In 1990 a Cornell University doctoral candidate, Bernie Tershy, undertook his thesis research on San Pedro Mártir Island in Baja, Mexico. Like the Farallones a major seabird destination, San Pedro Mártir's mother lode of guano had been exploited in the late 1880s by the Mexican Phosphate and Sulphur Company, and one wonders if Lucifer himself was sitting on the board of directors. In three years the company deployed well over one hundred Yaqui Indians to scrape the guano with pickaxes and shovels, removing one thousand tons of the stuff from San Pedro Mártir and a neighboring islet per month.[24] In the process they trampled thousands of nests and many birds flew the proverbial coop. The phosphate and sulfur folks gave as good as they got and left behind an invasive black rat.

Tershy set up camp to study the brown- and blue-footed boobies, and what he got to see was not pretty: rats persecuting birds, hopping up on their backs and chewing them alive. Tershy told me that he really got an emotional sense of the injustice of invasives when a rat bit his sleeping child. Tershy's frequent colleague Don Croll had watched similar devastation by rats on islands in Antarctica and the two of them conferred. They could keep their rears in the cool seat of science or they could "do something about this."

Only 5 percent of the earth's land mass is made up of islands, but 20 percent of all bird, reptile, and plant species live on them. Moreover, an overwhelming majority of extinction exceeding what would naturally occur absent human impacts occurs on islands. That's 95 percent of bird extinctions, 81 percent of reptile extinctions, 54 percent of mammal extinctions, and 62 percent of plant extinctions. Of 871 extinctions since 1500, 80 percent occurred on islands and at least 54 percent of those were directly due to invasions.[25] If you want to do something about stemming extinction, islands are a good place to start. Invasions on islands have an upside. Although the process involves the mortality of the invader, they can be completely expunged from islands in a relatively straightforward way. When island invasions are stopped, the damaged ecosystems frequently recover.

Biological restoration often has a cultural element as well. Island Conservation's most ambitious eradication to date occurred on what was known for hundreds of years as Rat Island in the Aleutian Island chain. Norway rats scurrying off a Japanese shipwreck in the late 1780s spent the next two centuries decimating the seabirds there, giving the island that nasty moniker. In 2008, Island Conservation and its partners eradicated the rodents. Within very short order, seabird numbers and variety began to increase, rock sandpipers, pigeon guillemots, red-faced cormorants, and black oystercatchers among them. And locals reclaimed the island in another way as well—it is now called Hawadax Island, from an Aleutian word meaning "welcome."

As newly hatching academics, however, Tershy and Croll were initially somewhat inhibited about going in and taking a species out of the system. "To interfere with nature was seen as poor science, if science at all," said Croll, adding that in the university setting, "you had to kind of hide that you were doing it."

Tershy and Croll were encouraged by one important academic figure, however: Michael Soulé. Soulé enjoys a place in history as the father of conservation biology, the discipline he founded to marry pure science with its application on the ground. (My book *The Spine of the Continent* profiles Soulé at length.) A professor at UC Santa Cruz in the mid-1990s, Soulé had just a few years earlier founded the Society for Conservation

Biology and authored its seminal textbooks. Wilson and MacArthur's island biogeographical theory had resounded in Soulé's ears. As human impacts are indeed making islands out of terrestrial landscapes, his all-too-prescient insight was that this would accelerate extinction. Soulé had taken a major step away from the world of academic evolutionary biology to assert that scientists are morally compelled to apply their knowledge to real-life problems. He supported Tershy and Croll, and in 1997 they founded their nonprofit, Island Conservation (IC).

"Not only was Soulé a board member when IC was founded, I had taken courses from him as an undergraduate and was hugely influenced by him as well," Brad Keitt, IC's director of conservation, told me. Soulé brought in guest speakers like the philosopher Arne Næss, who talked to the students about the concept he called "deep ecology," which asserts that nature has a right to its life outside of human utility. Keitt credited Soulé not only with giving a kind of older-generation stamp of approval to IC's work but also with pointing the way to a direction IC is more and more going in today, which is toward community collaboration. This is otherwise known as "public participation in scientific research" or "co-creation," but let's just call it citizen science.

"Soulé's point was, biologists can't do conservation. You have to include the social aspect of it. Biology and sociology and interdisciplinary work would train the next generation of conservation," Keitt said. "That doesn't mean it's all about humans. But you can't do conservation without addressing human communities and engaging in their issues. Most conservation NGOs are staffed by biologists who like to sit by themselves on islands and count birds. So we're trying now to get the scientists talking to people to better explain the issues at hand, to prevent extinctions, but to also find the leverage points and common ground where local communities get in on the act from the get-go."

Until recently, IC mostly operated on small, remote islands where no humans live. "We have been able to be more tactical, biological, about how we do the eradications," Heath Packard, IC's director of philanthropy and communications, told me. "But the fact of the matter is, given the thousands of species that are threatened and endangered on islands and trying to eke out an existence, there isn't enough time or resources to get to all those places. We're trying to get to where the extinction risk is greatest and where the return on investment is highest. With biodiversity that tends to be on larger islands that have more diversity and more invasions. The larger they are the more complex—and the bigger the social dimension."

To scale up its operations, IC is increasingly employing a strategy from the realm of social sciences and peace and conflict resolution. "It's all based on scientific principles and psychology," Packard explained. "Citizen science doesn't have to be limited to ecological sciences, but includes local values in cultural, historical, and economic connections to their islands. The process helps us see the values and objectives of the local community."

Keitt is leading the charge and told me how IC approached an eradication of cats from Natividad Island, again off the Mexican coast of Baja, where Keitt originally went to study the black-vented shearwater. Ninety-five percent of the world's population of this nocturnal seabird nests on Natividad, in the ground. "During the day you see the ground with holes in it, and you think giant rodents must live there," Keitt explained. "But at night the island comes alive with bizarre calls and wings whizzing over your head. I was there in the midnineties to study the species in the usual ways—to figure out a population estimate, reproductive ecology, that kind of thing." The island has been inhabited by people for years, and its forty-year-old town includes a desalination plant, a school, and a cannery.

"We knew there were cats on the island brought fifty years before to deal with native deer mice by fishermen who lived there. One of my goals was to measure the impact of the cats on the birds. Since they are nocturnal, many of the island inhabitants had only ever seen dead ones. We used video cameras with infrared to look down into the nests, and we showed the film to schoolkids. We explained the worldwide relevance of Natividad to this species of bird and explained that the cats were threatening them with extinction."

Keitt said that once the locals became aware of their treasure, and reflected on its uniqueness, community pride was immediately piqued. "The way folks outside Mexico looked at Natividad really resonated with them. They discussed the situation. Cats are killing one-of-a-kind birds that don't have any place to live besides here. Maybe we should get rid of the cats. The community itself came up with that decision. We hired local individuals to help with the removal." A partner organization based in Mexico "helped link the people's desire to the government agencies from which we got the permits."

Circling the Farallones aboard the *Fulmar* and pondering the official reluctance to let me actually step foot on them, I found myself agreeing with the general sentiment that the fewer people who trod among the wild creatures in this lonely place, the better. I had developed a highly romantic view of the scientific expedition, the brave going-forth into the unknown to discover and document, but I had to admit that the expedition narrative is itself double. For one thing, historical expeditions have spread invasive species around the globe. While it is dark indeed to call *Homo sapiens* the ultimate invasive species, usurping the evolutionary space of so many other beings, there is truth in the characterization. So what is a citizen to do? The horse is out of the starting gate, so to speak—with population numbers set to reach ten billion people on earth by 2050, no space will be exempt from potential human inhabitants.[26] The ocean darkened as the sun began to set, and with the day drawing to a close, the researchers aboard the *Fulmar* started to sit back, to show in laughing faces the elation of endless water, endless sky. Heading back toward my own home among the sparkling lights of the advancing city skyline, I was happy to leave the seabirds to theirs, in the deepening dark behind us.

CHAPTER THREE

The Wild Garden

While hunters avidly set to West Coast waters, depleting them of whales and sea otters, terrestrial California was spared the churning drives of empire until the late eighteenth century. But when the Spanish did hit California, the impact was seismic.

Motivated by news of the encroaching Russian fur trade, in 1769 the Spanish Crown directed Captain Gaspar de Portolá to follow up on the starry-eyed report of a perfect bay in California, made 167 years earlier by Sebastián Vizcaíno. "Discovering" Monterey Bay in 1602, Vizcaíno had come upon "the best port that could be desired," capacious and wind-sheltered.[1] "The land of this country is very fertile," reports the expedition diary, "and has good pastures and forests, and fine hunting and fowling. Among the animals there are large, fierce bears, and other animals called elks, from which they make elk-leather jackets.... On the beach was a dead whale and at night some bears came to feed on it."[2]

Portolá went looking once again for an imperial stronghold. Alan Brown, whose 2001 translation of Padre Juan Crespí's diaries of the Portolá expedition have deepened and revised how we understand the landscape today, writes that in looking for landmarks, the exploring party referred to Vizcaíno's testimony and several other sources "so constantly, in fact, that the descriptions became burned into the explorers' minds."[3] Vizcaíno had reported a bay large enough to hold the Spanish fleet. Looking for what was an exaggeration, and noting the small size of Monterey Harbor when he reached it, Portolá thought they must not have yet arrived at Vizcaíno's grand bay. He kept heading north.

By the time he reached the Whitehouse Creek watershed, about a mile inland from the coast at Punta del Año Nuevo, Portolá and his men were exhausted and running low on supplies. Crespí documented the occasion: "Here we stopped close to a large village of very well-behaved good heathens, who greeted us with loud cheers and rejoiced greatly at our coming." He further documented a "very large grass-roofed house, round like a half-orange," and wrote that it was large enough to contain the whole village.[4]

The men were fed and given native California tobacco. The Spaniards reciprocated by giving the Quiroste people domesticated tobacco—prefiguring the biotic exchange about to take place on a larger scale. Led by Indian guides, Portolá and his men continued up the coast and a few days later "discovered" San Francisco Bay from a ridge above present-day Pacifica.

"The Spaniards were lost, at the end of their rope," California State Parks archaeologist Mark Hylkema told me. "The Quiroste headman could have turned them away. Instead he decided to host them."

The Spaniards soon began to establish missions along the coast, disrupting an Indian way of life that had evolved with the California landscape for thousands of years. In short order the landscape began to be cleansed of its top predators, wolves and grizzly bears. Fire, an agent of transformation used by Native Californians to manage their food supplies and their cultural life, was looked upon by the Christians as holy on the one hand and demonic on the other. The Spanish imposed themselves on the landscape and on the people and understood neither. Burning was forbidden.

The landscape and the indigenous people went into a steep decline, and the impacts from these biotic and abiotic disappearances continue to have negative repercussions today. In 2015 I drove into Yosemite National Park and saw firsthand why Governor Jerry Brown declared a "tree emergency" here. Literally millions of dead trees stood red and desiccated on a drought-stricken landscape. If indigenous burning practices had not been interrupted lo these several hundred years, the state might not be in its current danger of literally going up in smoke. Among other things, low-burning intentional fires cleared out duff and debris and thus moderated the intensity of lightning-strike fires when they did occur.

Can't Bear It

An original citizen of California, the grizzly is now found here only in effigy and depicted on the state flag. Descended from a bear lineage that goes back twenty million years, the grizzly evolved approximately 1.5 million years ago, in the company of the Pleistocene megafauna. By comparison, the grizzlies were little guys and gals, their three- to eight-hundred-pound profiles dwarfed by the six-ton woolly mammoth, for example, or the elephant-sized giant ground sloth. These big, big ones began to disappear from North America at around the same time the first people arrived in the Pleistocene. The largest animals went first, over the course of a few thousand years, eventually leaving the grizzly to take the top tier in body mass. Historian Peter Alagona says, "By ten thousand years ago, they were the second-most-dominant land animals in California, after humans."[5] The word *dominant* is problematic—for one thing, the wolf was still in the neighborhood when we had grizzlies, and arguably had a bigger effect on the rest of the food chain—but for the moment, we know what he means.

One persistent apologia for the California conquest by Spain had it that the imperial power was actually doing the Indians a favor by dispatching grizzly bears with relative ease. Grizzly bears raided stores of grain and nuts that took months to collect. It is likely that any downed animal, like a deer or elk, obtained again by long labors, was fair game (literally) for any grizzly bear lurking around, waiting to push the hunters off their kill. But the grizzly was also woven into the worldview of the people, addressed by Indians continent-wide as "great-grandfather." This position of anteriority, procreation, and respect is shared in European associations with bears as well.

Today, we can articulate another framework around the role of the top predator grizzly bear in historical California. There were more grizzlies in California than anywhere else on the continent (with the exception of what is now Alaska) for the same reason there were among the highest population densities of indigenous peoples here anywhere north of Mexico: the fantastic panoply of different kinds of terrains upon which to make shelter throughout an annual cycle of varied but overall temperate climate, and the attendant abundance of food sources. They contributed to the functioning of the ecosystem in myriad ways. Their life cycle was deeply twined with that of salmon, the fish we most associate with their diet. Throwing salmon carcasses willy-nilly all over the place, grizzlies provided fertilizer for riverbanks and food sources for smaller mammals, invertebrates, birds, and microbes; they created a bridge between aquatic and terrestrial nutrient cycling. Digging the ground for tubers and bulbs, they aerated the soil and stimulated the growth of plants— just like we do when we're digging up bulbs in our gardens. They dispersed nutrients from berries and nuts they consumed, and in defecating at a different location from where they ate them, helped dispersal. On the California coast, grizzlies chowed down on beached marine life, and probably aggressively dove in to nab seal pups, thus helping to cycle ocean nutrients inland.

Grizzlies get close to top billing in the trophic cycle, like sea stars do, but their generalist ways distribute their effects. To remind you, the trophic, or food cycle, describes the direction of impacts between the top of the food chain and the bottom. The sea stars wasting away in the first chapter of this book are top predators in the intertidal, controlling, for example, sea urchin populations and preventing those spiky critters from decimating the kelp. Now that so many sea stars are among the disappeared, the sea urchins are indeed devouring the kelp. In terrestrial North America, scientific focus has landed more frequently on the wolf as the kingpin in the cycle, though grizzly bears are often considered right along with them.

In sum, life is better for grizzlies when wolves are around, because wolves keep the population of herbivores under control, which thus preserves more berries for the bears; grizzlies also take advantage of the wolf's hunting prowess and frequently push wolves off a kill. The wolves don't mess with the grizzlies, and wait until the bear leaves the carcass before having at it again. So to my quibble with calling grizzlies and humans the "dominant" life-forms after the disappearance of the megafauna: number one, there were wolves in California at the time so let's not leave the valedictorian out of the yearbook; number two, plant ecologists and entomologists, to say nothing of microbiologists, will be clearing their throats and saying "excuse me" about the relative primacy of their own beloveds in the way ecosystems work. At the same time, I'm bothering with this at such length because carnivores with the biggest body masses are going extinct at the highest rate today and the impact of losing them penetrates to all layers of biotic and even abiotic interactions.

Grizzlies still roam parts of the lower forty-eight in Washington State, Idaho, Wyoming, and Montana. Citizen science efforts to help them keep on keeping on include a program to monitor camera arrays with a nonprofit called Conservation Northwest. Remote cameras keeping tabs on where they are ironically help reduce the biggest threat to grizzly health, which is proximity to people. Check out this Flickr stream of photo-captured bears and their big-toothed brethren: flickr.com/photos/conservationnw. We Californians have to content ourselves with helping our remaining top carnivore, the mountain lion, make it from here to there. Citizen scientists working with De Anza College instructor Julie Phillips helped identify Coyote Valley in the heart of Silicon Valley as a wildlife corridor, documenting among other wildlife movement the comings and goings of mountain lions. They got "safe passage" in Coyote Valley until the landscape comes up for review again in 2040.

All Consuming

Over a range of ecosystems from terrestrial to aquatic and all over the globe, the top-down impact in the trophic cascade is made by a carnivore. This is what trophic cascade biologist Robert Paine showed with the sea stars in the tide pool. Biologist James Estes did equally well-known work testing the idea in deep open waters, with sea otters. They figured out that when these top predators were removed from their environments, the ecosystems they were part of fell apart. But think about trying to figure this out with terrestrial top predators—like the jaguar, for example. Not easy.

Thus when in 1990 John Terborgh found himself thinking hard about why and how terrestrial predators go extinct at faster rates in fragmented habitat, he had no way of actually testing his ideas until a friend told him about Lago Guri, a man-made lake in eastern Venezuela. Terborgh's thinking started with island biogeographic theory.[6] IBT, again, is the scientific template provided by E. O. Wilson and Robert MacArthur that allows conservation biology to study how isolating pieces of landscape leads to higher rates of extinction.

Terborgh is one of the world's preeminent tropical rainforest ecologists. He founded and directs the Center for Tropical Conservation at Duke University's Nicholas School of the Environment. Primarily interested in the ecological interactions that create diversity, he was initially drawn to the tropics because here "there are still intact ecosystems. Everything is massively disturbed in North America. I wanted to study how nature operates independent of the interventions of human beings." In fact his most well-known work is based on a massive human intervention, a dam built at the confluence of the Orinoco and Caroni Rivers in east Venezuela in 1986. The Guri Dam created an artificial lake fifty miles wide and five hundred feet deep; the highest peaks in the countryside poked above the water, instantly creating islands of different shapes and sizes. The created islands would pose an opportunity to monitor extinction as it unfolded on them, according to the precepts of the island biogeographic theory, which provides that the smaller and more isolated an island, the fewer the species you will find on it.

Terborgh was convinced that island extinctions don't just happen via passive processes, like random fluctuations in numbers, but are also driven by ecological processes, particularly those instigated by big-toothed carnivores. "I remember when I first became aware of how important predation is," he told me. "In the early 1980s I collaborated with Louise Emmons, a really tough field person," and currently an adjunct research scientist at the Smithsonian. "We set up a project to study jaguars in Peru. The results of that study forced an understanding, a recognition on my part, that what these very rare animals are doing has a huge impact on everything else that goes on, by regulating prey populations."

At the time, he said, "The popular idea was that predators were not taking down healthy adult animals but the weak and the sick, harvesting nonessential, disposable if you will, elements of the population; therefore, not exercising control over the population itself. That idea came out of rumor and misinformation, maybe more than anything else because of the scarcity of these animals. If you've seen a mountain lion in the US, you've seen something most people don't. They're there, and they have a very powerful effect. But they're super good at being invisible."

Terborgh and Emmons radio-collared jaguars to keep tabs on their movements and collected scat to see what they were eating. Their data led them to construct a diagram of the animals' effect on the species farther down the food chain. Jaguar prey include mammals like peccaries (piglike) and capybara (a large rodent), and reptiles including snakes, lizards, turtles, and crocodiles. "And the prey do all kinds of things to the plants, threatening them by eating lots of seeds, and benefitting them by swallowing those seeds and moving them to other places. All these processes structure how the whole collective works."

The creation of the islands at Lago Guri provided Terborgh with a ready-made site to study the effects on the ecosystem of removing the top predator. "Although it was an obvious experiment—remove all the predators and see what happens to the prey—it had never been done. Lago Guri constituted a fortuitous, multibillion dollar experiment"—one far beyond the reach of scientists, with their perennially limited budgets, to create. Terborgh went to Lago Guri in 1990; in the four years since the islands had been created, the carnivores at the top of the region's food chain (including jaguars) were already gone. Even the largest landscape remnants did not have enough prey to support a top predator. As described by the title of his 2001 *Science* paper on the subject, "Ecological Meltdown in Predator-Free Forest Fragments," what he found was a nightmare world of consequence unfolding among the remnant species on the islands.

At Lago Guri, Terborgh and an international team of scientists ascertained that while there were no top predators on the man-made islands, the next level of the food chain, the so-called mesopredators (*meso* means "middle"), had increased in number on medium-sized islands. Now unchecked by jaguars, for example, capuchin monkeys proceeded to devastate bird populations through nest predation. Herbivores similarly boomed in unhappy numbers. Howler monkeys that historically lived in groups of three to seven were now found in numbers thirty times as high. They were also starving. In response to so many monkeys eating the greenery, besieged trees laced their leaves with toxins; the monkeys ate the leaves and vomited. Leaf-cutter ants that chew greenery to make fungus farms reached populations one hundred times higher than their mainland densities; their exponential impact denuded the forest, eventually killing the trees. Thus from the loss of the cryptic carnivore, the entire food web basically collapsed.

"What John did was take Bob Paine's keystone species idea into the terrestrial field and change the scale," Justin Brashares, a professor at UC Berkeley, told me. Terborgh was perhaps the first to associate the ecological impacts of species interactions with the size and placement of habitat. Brashares is part of what is now a second or even a third generation of scientists looking at trophic cascades, and he is an author, with Paine, Terborgh, and many top-tier researchers, of the 2011 *Science* paper "Trophic Downgrading of Planet Earth." Jim Estes is the primary author, and he notes that species interactions can be invisible until perturbed, and that the loss or the addition of a species can take years or even decades to become apparent. He also notes that "populations of large apex consumers have long been reduced or extirpated from much of the world."

Brashares updated me on some of the pushback directed at the concept of top-down forcing. "There are those who hold that the system works from the bottom up," he said. "I am willing to accept that top-down doesn't regulate every ecosystem everywhere. Some communities may be regulated bottom up. But today trophic science is looking with a more nuanced view."

Brashares devised a study that takes at its starting point neither the top nor the bottom of the food chain, but the middle. "We are looking at what happens when hippos are hunted out of the ecosystem, or when dramatic changes in rainfall or agriculture impact water levels and cause hippos to disappear." Hippos are herbivores. "They are amazing ecosystem engineers," Brashares said. "They go out each night and eat all these nutrient-rich grasses, and then defecate everything into their water. Rich, diverse aquatic communities result. There are all sorts of ecological consequences when the hippos are gone. You have a dramatic change in the aquatic system, which affects vertebrate communities from birds to otters."

In other words, starting with an impact from the middle, both the bottom and the top of the whole food chain are affected by change in this interaction. As Estes and colleagues write: "The topology of ecosystem dynamics is now understood to be nonlinear and convoluted." And subsequent levels of impact can lead to "tipping points (also known as thresholds or breakpoints), around which abrupt changes in ecosystem structure and function . . . occur." Elsewhere in the article, he writes, "Connectivity . . . holds that ecosystems are built around interaction webs within which every species potentially can influence many other species." Bottom, top, and middle: it's all connected.

Querying Jim Estes about the impacts of trophic cascades, I asked him a simple question. We do without top predators in a lot of places—in California we still have mountain lions, but no grizzlies or wolves, which historically were here. (And the historical range of the jaguar was as far north as Monterey.) "Things seem to work okay without them?"

Estes considered my question with a certain freighted forbearance, putting me in mind of poet and feminist Adrienne Rich's utterance, "A wild patience has taken me this far." "I would respond to this question by challenging the legitimacy of the claim that things seem to work okay without them," he told me calmly. "I suspect that in many cases this is a false perception, based largely on the facts that we only 'know' systems without the big predators and we have no clear basis for judging what their losses might have meant. For example, it is becoming increasingly clear that the Lyme disease epidemics in the midwestern and northeastern United States are linked to changes in coyote populations. So are midwestern and northeastern ecosystems working okay? I've often wondered if California's chaparral wildfires were as intense in times past when the full complement of Pleistocene megaherbivores was roaming the landscape. These are issues we are just beginning to think about and understand."

Estes also commented, "Many of the things we have done to the earth might be remediated, at least to some degree. Species loss, however, is not one of these."

Light Eaters

Another perspective on the reverberating impacts of taking large-bodied mammals off the face of the earth is described by paleontologist Anthony Barnosky in his 2014 book, *Dodging Extinction: Power, Food, Money, and the Future of Life on Earth*. Paleontologists, of course, dig up fossils and study both the objects and the strata in which they were found, and put together the story of life as it has unfolded on Earth over time. They are the ones analyzing how many megafauna lived when and where, and when they disappeared. Barnosky calculates that historically, 350 species of megafauna could have been supported on Earth at the same time, based on the amount of energy produced by plants and algae.[7] Those are the primary producers, the leafy green stuff that is turning energy from the sun into sugar and carbohydrates—embodying energy, which is then transferred to the organisms that feed on it.

Every time one animal eats another, there is a net loss of energy in the transaction. So you start out with a certain amount of available energy from the sun, and after plants use it to grow, they contain a bit less energy than the original power of light. Then herbivores eat the plants, and now *they* embody energy that came originally from the sun, and a certain amount of interest is taken off that original power number once again. So you can see that by the time we get to the megafauna at the top of the ancient food chain, there's only a certain amount of energy left to create and sustain their big bodies. Barnosky explains, "It is possible to estimate the body size of an extinct species, how many animals there were on average in a square mile (or kilometer), and how many square miles (or kilometers) the entire geographic range of the species covered." Then you can add up the total weight of all the individuals of all megafauna species therein.

To talk about how much energy is produced by the sun and then harnessed by photosynthesis, Barnosky trots out a word that I frankly had to spend some time contemplating on Wikipedia: *exajoule*.[8] This is a unit of power named for physicist James Prescott Joule, and he got quite a lot more kinds of joules named after him as well. Beginning with the nanojoule, which Wikipedia helpfully says is "1/160 of the kinetic energy of a flying mosquito," we move on to microjoule, millijoule, kilojoule, megajoule, and gigajoule before getting to exajoule, which is one quintillion joules. Now we are in Aladdin's cave, with all those joules. It's easier for me to think of an exajoule as a ten-to-the-eighteenth-power joule, I don't know why. Wikipedia says the 2011 Tōhoku earthquake and tsunami in Japan had 1.41 exajoules of energy in its 9.0 "on the moment magnitude," which is a phrase Hemingway might like. There are more joules, too, including the yottajoule—they all pretty much sound like bad jokes. The thermal output of the sun is four hundred yottajoules per day.

By the time the sun's energy is converted by photosynthesis, there are approximately 728 exajoules available per year, which, Barnosky said, "have been more than enough to power the global ecosystem on land, and all the species in it, before humans became so abundant on the planet." We know it's been more than enough because for millions of years, some energy that could have been available for powering on died and decomposed, eventually becoming coal in the deep layers of terrestrial earth and oil in the deep underground of the oceans. Barnosky says that megafauna biomass stayed about the same for hundreds of thousands of years on earth, and then with the Pleistocene die-off ten thousand years ago, dipped way down. By about three hundred years ago, the biomass was back up to pre-die-off levels, but "instead of all that weight being distributed through many big-bodied species, an enormously large proportion of it was made up of human bodies and the livestock we breed." We took over the predominance of energy available to make biomass into other life-forms and we made it into ourselves, and cows. This sounds like some horrible science fiction scenario—people and cows replacing every woolly mammoth and saber-toothed cat, and now well on their way to replacing every lion, tiger, bear, and whale. If only it *was* a bad dream.

"Then something remarkable happened," Barnosky relentlessly continues. "Megafauna biomass suddenly began to sky-rocket, in just three centuries rising to the level we see today—to nearly one-and-a-half *billion* tons—orders of magnitude higher than it was for all of the time humans have been on Earth. About one-fourth of the 'extra' biomass is us; the other three-quarters are our livestock." How did we do it? By breaking through the photosynthesis barrier. By using more than the sun has to offer. The timing is coeval with the industrial revolution, by which we learned to dig up and drill up those reserves of photosynthesis past, and burn it up. *Finis*: "The human race currently needs more energy just for itself than has normally been available to power the entire terrestrial ecosystem."

Barnosky is one of the scientists I habitually trail around, so he knows me. "Congratulations, Tony," I emailed him upon reading his book, "you are better at telling a horror story than Stephen King. The hair is standing up on my neck."

He wrote back that his work is "meant to be hopeful," because it's quite possible for us to get off fossil fuels and to thus stem extinction. "The key perspective," he said, "is that it's essential to integrate other species into human-dominated systems—because that now describes the condition of most of the planet." It's also essential to reintegrate that gift from the gods we have so parsimoniously withheld from our ecosystems: fire.

The Burning Bush
In July 2015 I again joined Rob Cuthrell, then into his second year as a PhD archaeologist, and members of the Amah Mutsun Native Stewardship Corps in the Santa Cruz Mountains. They had been camping for weeks, but the dusty Cuthrell somehow managed to maintain his fresh demeanor. We waited while the chairman of the Amah Mutsun, Valentin Lopez, led members of the corps into a shielded green space to smudge the site and to pray. Smudging is a ritual that involves burning dried sage, one of the culturally important plants historically grown here. Cuthrell walked in advance of our vehicles on the small dirt road branching off from a larger one that snakes up the mountain. The California Department of Fish and Wildlife had recently directed this protocol to prevent running over red-legged frogs and San Francisco garter snakes, both endangered species. It felt like a counterpoint to the Mutsun ritual, a way to viscerally enter the place with respect.

The workday commenced. What might have looked to a passing hiker like any weed-whacking crew was part of a model effort to integrate what is sometimes called "traditional ecological knowledge" into current-day natural resource management. Whooshing machetes opened up new spaces of sky as towering stalks of poison hemlock were felled. The hemlock has been crowding out a potentially healthy population of tarweed, a member of the sunflower family with ethnobotanical significance for the Mutsun. Historically, many Native Californians burned the still-green tarweed near the end of the summer. The scorched mature seeds could thus be crushed and mixed into pinole, a porridge-like staple of coastal diets. These plants have grown here for thousands of years.

Lopez and I stood and talked while the stewards worked. Lopez is tall and gently authoritative. His right eye drifts as if keeping tabs on another dimension. He has more than once reminded me that today's California Indians live in two worlds. They are pretty much always tacking back and forth between past and present, between traditional and contemporary modes. On this landscape, the double vision includes native plants that have persisted here despite centuries of invasions and land-use conversion, a twined history they are helping to parse.

Dividing Line

The landscape at Quiroste was initially recognized for its historical significance by archaeologist Mark Hylkema. Logged, ranched, and farmed for decades, the property was donated to the parks in the early 1980s. Hylkema had a bee in his bonnet from reading expedition-based documents of Spanish encounters along the coast, and set about locating the remains of what Crespí called Casa Grande, the large roundhouse structure where the Quiroste hosted the Spaniards. Hylkema was motivated because, as he put it, "Here is where prehistory became history." Today, the site is frequently referred to as Metenne, the name of one of several villages that were part of Quiroste territory. Working with Crespí's expedition diaries, Hylkema eventually located multiple sites on the valley floor of Whitehouse Creek that evidenced ancient Indian presence. "That's a lot in one place," he told me, so he zeroed in. With the help of students from Cabrillo College, in 2002 Hylkema radiocarbon-dated building material and plant remains, which showed that one of the major sites in Quiroste Valley was occupied at the onset of Spanish colonization. The site and environs were thus protected as part of a 225-acre Quiroste Valley Cultural Preserve within Año Nuevo State Park. Hylkema turned to Chuck Striplen, then a UC Berkeley doctoral candidate and a member of the Amah Mutsun Tribal Band, to help explore the site. While there are no known living descendants of the Quiroste people, the Amah Mutsun trace their lineage to historical polities in the area. Striplen, now an environmental scientist and a visiting scholar at UC Berkeley, assembled a multidisciplinary team, including UC Berkeley archaeologist and California Indian scholar Kent Lightfoot, to begin on-site investigation. The ongoing collaborative research at Quiroste includes the Amah Mutsun, researchers from disciplines such as ethnobotany and fire ecology, and state agency natural resource management professionals.

This is "extreme citizen science" in that traditional Western archaeology is being expressly "co-created" both to answer academic questions and to facilitate present-day goals for local people.[9] The archaeologists are figuring out the lifeways of precontact indigenous people, and studying their impact on the ecosystem. The Amah Mutsun are after something more. Lopez recounts that in 2005 tribal elders came to him and said, "We have to get back to taking care of the land. Creator never rescinded our obligation to it." Lopez laughed. "Can you imagine, these people with minimum-wage jobs if they have them at all, who don't own any land themselves, saying we have to steward it? Now where were we going to do that?" He added that saying no to tribal elders was not an option. When Hylkema and Striplen invited the tribe to participate at the Quiroste site, Lopez saw an opportunity.

Details, Details

In 1770 Junípero Serra took up residence and his position as

"Father Presidente" of the Alta California missions (differentiated from the Baja missions). The phrase *missionary zeal* describes his lust for harsh piety. Many Europeans still equated California with Eden, but for the place to actually suit their outlines of paradise, it had to be cleansed of top predators—good-bye, grizzlies and wolves—and quenched of fire, and its population of heathens had to be baptized. Kent Lightfoot calls the mission complexes "massive enculturation machines—monumental agricultural and craft centers" where the padres "would attempt to transform California Indians into Catholic peasants." Serra never wavered in his belief that he was doing the right thing by "saving" the Indians, and according to Alan Brown approached the missionary career as "heroic."[10] In addition to his passionate emotionality, Serra was analytic and strategic. He considered Crespí's expedition journal to be a critical tool in marketing the mission system to the Spanish back home.

Journalists who bridle at editorial reins may sympathize with Padre Juan Crespí. Serra considered Crespí's fastidious expedition notes overwritten and suggested he trim the "minutiae." Today Crespí's minutiae bear priceless testimony to California precontact: he described plentiful bears, redwood forests that seemed to go on forever, fields carpeted with wildflowers, and fires burning as Portolá's expedition traveled up the coast. Crespí responded to his boss: "Then you do not want me to tell things just as they are, or as I saw them?"[11] Serra eventually changed his tune, and three days before he died let his biographer know that he wanted Crespí's work extolled. Serra implored greater distribution of the expedition diaries to Spanish readers who would "be inspired to come here! Merely reading the journals would suffice to move no few of them to leave homeland and native province, and take up the voyage to come and labor in this vineyard of the Lord."[12]

In the sixty years of its existence, the mission system did profound damage to the indigenous California way of life and the biodiversity the Indians cultivated. Suppressing California Native burning practices and inserting grazing and farming on the landscape further suppressed native plants and invited invasives to take over. Like invasive rodents, invasive plants are transplants to areas where they did not coevolve with predators and so outcompete the natives. Typically getting hold of a disturbed landscape—where a road has been cut, for example, or where land has been converted to agricultural use—they crowd out native plants. And so go the native invertebrates like butterflies, and here we go through the food chain again, like removing steps on a ladder: good-bye, native reptiles, amphibians, birds, and mammals.

The Indians lost their customary food supply and medicines. Many were forced into the mission system for simple survival; their labor on behalf of the Spaniards was largely coerced. The Native population was further assaulted by exposure to diseases brought by Europeans who had developed immunities to them. Records from the San Juan Bautista Mission (near Hollister), to which many Amah Mutsun trace their ancestry, document 3,396 baptisms, 848 marriages, and 19,421 deaths between 1797 and 1833.[13]

Mexico won independence from Spain in 1821 and, as it took political control, directed secularization of the missions. Land ownership cohered around large rancheros, and many of the mission Indians went to work on them. One outcome of this period is that original tribal identities and territories were further occluded, as people moved around to survive. Many Amah Mutsun people settled in San Juan Bautista. The infamous gold rush began in 1848, multiplying the European American population here and speeding the commodification of what in 1850 became the state of California. In a frenzy of moneymaking and environmental degradation, natural resources were grabbed up and laid claim to. Indians with ancestral rights to the land were now impediments to American Manifest Destiny. On January 7, 1851, California governor Peter H. Burnett delivered an annual message to the legislature, synopsizing a violent impasse between the white settlers and the Indians. "We have suddenly spread ourselves over the country in every direction," he reported, "and appropriated whatever portion of it we pleased to ourselves, without their consent, and without compensation." He asserted, "That a war of extermination will continue to be waged between the races until the Indian race becomes extinct, must be expected. While we cannot anticipate this result but with painful regret, the inevitable destiny of the race is beyond the power or wisdom of man to avert."[14]

Homeless at Home

"The first five years of my life I spent living in tents," Lopez had told me on an earlier occasion. "Our people were very poor. We did the crops and sheep shearing and moved from ranch to ranch. Some of those land owners in Hollister and Gilroy knew that we were the Natives of the missions, so they allowed us to set up tents on their ranches when we didn't have a place to live."

Lopez was born in 1952 and grew up in Morgan Hill, about seventy miles south of San Francisco. When he went to school, "We were to say we were Spanish or Mexican but not Native American. As children we didn't understand that, but we knew there was an important reason behind it so we obeyed." Lopez's grandmother's grandmother had told her she remembered indentured servitude in the missions, the official order to exterminate American Indians, and the practice of kidnapping children and placing them in orphanages.

As Joseph Campbell might put it, Chairman Valentin Lopez is a specially achieved individual. Frequently first on the docket of Bay Area conferences that have to do with wilderness, nature, or open space, he inaugurates these proceedings with a prayer. It is a sight—plaid-shirted, Carhartt-trousered biologists with heads bowed. His life path illustrates the hero's journey par excellence, though it does not entail leaving home. Lopez's quest has to do with recuperating knowledge, practice, relationship, community, and identity right in the place where these were lost. Lopez's dilemma was that since the Amah Mutsun do not have federal recognition, they have no tribal territory upon which to exercise their responsibility to the creation by caring for the environment. This is an existential quandary and a practical problem. Identity may be a birthright but it is also daily established through observances and actions on behalf of the natural world.

The Amah Mutsun have found multiple creative solutions to their challenge. Lopez has helped cultivate relationships with many local, state, and federal entities, including Pinnacles National Monument, the California Department of Parks and Recreation, the US Bureau of Land Management, and others. The Amah Mutsun have initiated restoration projects at Pinnacles, where they participate in the condor recovery program. (The condor is sacred to many indigenous people; it is a bird strong enough to carry the dead to the afterlife.) In 2011, the Amah Mutsun conducted a controlled burn there.

In 2009 Lopez and Striplen conferred with Rick Flores, curator at the University of California Santa Cruz Arboretum. Lopez told Flores that "tribal members had lost a lot of knowledge pertaining to plant uses and management," and they agreed to work together to re-create forty acres that replicate historical coastal ecosystem zones. Both tribal people and the general public are invited to participate and learn here. Flores told me he sees the work "as a form of environmental justice and a way to heal from colonization. This work is also about the sociocultural resilience of the Amah Mutsun and serves as a counterpoint to the discourse of historical trauma. They are still here." Becoming land stewards once again, Flores said, they are increasingly taking a place in "contemporary environmental planning and decision-making processes" from which they have historically been excluded.

At nearby Pie Ranch in Pescadero, cofounders Nancy Vail and Jered Lawson have provided the Amah Mutsun with space to cultivate a native garden. Vail and Lawson were "compelled to understand what it means to live on stolen land, how we can be in solidarity with Native peoples, and what our collective responsibility is going forward." Vail told me she is very much interested in "demonstrating how traditional land stewardship and modern organic agricultural techniques can be woven together," adding, "It's all a work in progress."

With the nation's first nonprofit devoted to nature protection, the Sempervirens Fund, Lopez has established the Amah Mutsun Land Trust, one of the first of its kind in California. And of course there is the work with UC Berkeley, to analyze and interpret the Quiroste site while simultaneously nurturing the area back to health.

While people without PhDs are contributing here, at the moment mostly by helping to remove invasive species, the project exemplifies "extreme citizen science" because its parameters revise the traditional frameworks around which archaeology and even conservation biology are undertaken. As the research is revealing the details of how human land use and habitation have driven historical ecosystem patterns and processes, it is also shedding light on the ways academic viewpoints have reflected political agendas. In the case of California Indians, academic blindness to their agency on the land is hard to extract from the historical drive to separate them from that land. As today's California Indians reinstate traditional natural resource practices, they also reassert a place for the human in nature that is positive—"co-creative" in helping to increase, and not deplete, wild species.

The Gift That Keeps Giving

California Indians pruned, coppiced, sowed, and weeded to intervene in the life cycle of plants and animals and to direct their growth and reproduction. But fire was their central management tool. Indians deployed a complex system of low-intensity staggered burns to provoke new growth of plants for food, medicine, and material goods, including baskets, canoes, and housing structures. They maintained different areas at different stages of succession. Newly burned areas attracted specialist bird species; freshly sprouted fields drew deer and other hunted animals within striking distance. Shrubby growth appearing a year or so after a burn was home to small mammals that were hunted and trapped. Mature forests were kept healthy with low-intensity burning of duff and other natural debris.

Multiple strands of evidence from the archaeological work at Quiroste point to the conservative conclusion that burning was practiced on this landscape for at least a thousand years prior to Spanish colonization. And that the fire regime had a significant impact on "the structure, diversity, and vitality of local terrestrial communities," according to an article by Lightfoot and others.[15] Archaeobotanical results show a high proportion of woody plants that can cope with low-intensity landscape fire, including redwood, California lilac, and alder. Lightning-strike fire is relatively rare on Northern California coasts, so the evidence points to frequent human-instigated burning. High proportions of grassland-associated food-plant remains are a further indication of intentional burning and contrast with the scrub and Douglas fir woodlands that characterize the valley today. Analysis of the density of microscopic glass-like structures called phytoliths that are left behind by some species of grasses also points to more of a grassland ecosystem than is found today, with fire as the likely mechanism for keeping shrub and tree encroachment at bay. And wind-borne charcoal particles from wetland sediments provide direct evidence of landscape fires.

The Indians carefully calibrated where and when to burn. Reproductive strategies of native perennial wildflowers evolved with fire; their seeds were important food sources. Hot flames ruptured seed casings and so promoted vegetation growth. The smoke from fire not only conveyed heartfelt prayers to Creator, but brought new life to some species that germinate as briefly as one minute after exposure to its nitrogen dioxide. Fire is intimately connected with water. Especially relevant to the state of our landscape today is the way that burning sustains the health and the breadth of wetlands. Burning around wetlands creates more edge habitat for more species, boosting nutrients that percolate through the hydrological carbon cycle to support fish, amphibians, and invertebrates. Wetlands act as sponges and filters of contaminants in the watershed, which are sequestered in the vegetation. Burning the vegetation and stimulating new growth can renew a wetland's capacity to clean the water. And fire keeps meadows open, enhancing their capacity for holding water and staving off drought.

The Christian view has appreciated fire as an agent of transformation, assigning its qualities of purification and flame to the Holy Spirit. But the Spanish and eventually the Americans did not see that this metaphorical formula has an organic inception.

Fire plays a central role in the geological carbon cycle, breaking down the woody and leafy products of photosynthesis into oxygen and nutrient-rich ash. In California and other dry places around the world, fire is the primary driver of decomposition. To stop fire is to stop what looks like destruction but is part of a cycle of regeneration. Fire reboots the system of life. As California Natives applied this central force of ignition with analysis and design, so they took responsibility for the human role in sustaining creation.

Twined with its centrality as a landscape-management tool is the significance of fire as the nexus of the California Indian cosmogony. "Hummingbird brought fire to our tribe," Lopez told me. "It was a gift from Creator." The Mutsun creation story recounts that after Eagle, Raven, and Hawk failed to, Hummingbird successfully wrested fire from underground badger people. This most tenacious of birds secreted an ember that flamed, and the result flashes in the bird's throat to this day. "We use fire in all our ceremonies, in our sweats, offerings, and prayers. And we have fires that are dances—we dance around the fire," Lopez explained. "The smoke carries what is in our hearts and minds, our prayers, to Creator, and is sacred for that reason." Significantly, fire is not just a gift received, but one that must be given back to nature as well. Lopez has repeatedly reminded me that while the science being discerned at Quiroste is all well and good, the real project here is about "stewardship."[16]

The Quiroste work has big implications for how we understand the role of humans in shaping the ecosystem here. It also has relevance for understanding the Anthropocene. Climate scientists predicting impacts from temperature and precipitation changes use models of human-driven land-cover change "that in general don't incorporate our knowledge of human history and how people used the land," according to University of

Chicago anthropologist Kathleen Morrison. Morrison told me that most of the literature on the Anthropocene doesn't "look at anything further back than two hundred to three hundred years ago," but humans have been changing the landscape for thousands of years. Morrison is spearheading the mother of all spreadsheets, an academic project called LandCover 6K, which aims to reconstruct land cover and land use back into what Mark Hylkema referred to as "prehistory," or the long time frame in which indigenous people impacted the structure of the landscape and which has been pretty much ignored by those outside the fields of archaeology and anthropology until now. Global change is coming about not just because we are burning fossil fuels, but also because of aggregating impacts from burning, cutting down forests, farming, and a myriad of other land use practices that impact the climate by way of biogeochemical and biogeophysical processes over time. The Quiroste work potentially bears on Morrison's project because it quantifies land-cover changes that resulted in altering those biogeochemical and biogeophysical cycles.

Parallel History

The Spanish, and eventually the Americans who followed, looked on what many considered an Eden and did not notice that the paradise here was human-made. There is no "natural" world for the Indians to return to but rather a highly managed landscape maintained by techniques to be recovered and reapplied. The history of plants, animals, and whole landscapes is often ignored or forgotten even by those who seek to care for these today. But successful restoration of ecological functioning in California requires careful understanding of how people and landscapes have been intertwined—and cast asunder—over vast periods of time.

The gaps in our understanding of how the ecosystem functioned precontact is not a problem for science alone, because it is nested in generations of cultural assumption. One perhaps surprising example is provided by that literary lowrider, American novelist Henry Miller, who takes as his reference point the Christian Garden of Eden to celebrate place in *Big Sur and the Oranges of Hieronymus Bosch*.[17] The 1957 title references Bosch's painting *The Millennium*, depicting a raving paradise: "[Big Sur] is the California that men dreamed of years ago, this is the Pacific that Balboa looked out on from the Peak of Darien, this is the face of the earth as the Creator intended it to look." Shortly dismissing any pre-Christian claim to the place by the Esselen Indians whose homeland he extols, Miller pays forward the prejudiced idea that this Eden, the very center of creation, is a European discovery. It has no past prior to Balboa.[18]

In terms of not seeing, even the naturalist John Muir sought to remove the sight of Native Americans like a mote from his cosmic eyeball. Muir observed Miwok people in the Sierra Nevada in 1869, noting an old Indian woman dressed in calico rags. "Had she been clad in fur, or cloth woven of grass or shreddy bark . . . she might then have seemed a rightful part of wilderness; like a good wolf at least, or bear." As Kat Anderson, whose magnum opus, *Tending the Wild: Native American Knowledge and the Management of California's Natural Resources*, has been a classic since it was first published in 2005, puts it, he was "unable to fit them into his worldview." With that attitude he helped to construct a philosophy of human-free wilderness—the enforcement of which was already degrading the ecosystems he loved to serenade. Muir may have known well how to read the rocks of Yosemite, but he did not understand that the wildflowers he found enrapturing were there largely as a result of active Indian management. "To both use and conserve nature,"

Anderson avers, "requires complex knowledge and practices, far more complex than leaving nature alone."[19]

World Fire

Indigenous burning practices are not limited to California, though every region, time period, and tribe has a unique story to tell. Anthropologists Doug Bird and Rebecca Bliege Bird are investigating common histories and contemporary experiences with fire among California Indians and Aboriginal Australians, such as the Martu people with whom the Birds have lived and worked over the past twenty years. The Birds collect data for their own academic purposes, and the Martu and other tribal people use that same data to help establish and defend territorial rights—these mutual goals make the project a "co-created" one.

Species, as has been noted, are going extinct all over the globe at a rate and magnitude one thousand times faster and bigger than "normal," and they are going extinct fastest in Australia. Australia has experienced the same loss of the top predator we have going on in North America; to review the drill, big-toothed mammals like dingoes (wild dogs) in Australia have a forcing effect on the rest of the food web, and when you take them out of the picture, the herbivores become overentitled to greenery and decimate it. Hosts of smaller species that depend on healthy vegetation start to blink out. Invasive species get a green light to come on into the ecosystem and start accomplishing their own outcompeting of natives. The view from North American conservation biology is that increased human use of the land, mainly by ranchers and farmers who actively remove predators to safeguard their livestock, is knocking the trophic kingpin off the pyramid and instigating a downward cascade of ecological disaster. But there are some interesting complexities in the Australian situation. The areas of the country with the least amount of ranching and agriculture—the least human impact—are experiencing the highest rates of extinction. In the central and western Australian deserts, moreover, endemic mammal losses are highest, but the dingo population hasn't changed. And where the Martu live and burn, species extinctions are in fact fewer and slower than elsewhere. When I saw Doug Bird's illustration of how this comes to be, my jaw dropped.

Most scientists' diagram of a trophic cascade starts with a top animal predator, depending on the ecosystem—a sea otter, or a wolf. At the top of Bird's PowerPoint slide: a Martu tribal member about to start a fire. "That's a person," I said. I made him laugh. "I can never understand why conservation biologists leave humans out of the picture," he said. Martu cultivate "seral stages" of vegetative growth after fire. In the first stage, newly burned ground exposes monitor lizards, which they hunt and eat; the second stage brings young sprouts of vegetation after the first summer rains, which produce seeds and fruits as the grasses mature, also harvested by Martu, and become fresh forage for other species. The grasses die, and spinifex, the so-called climax vegetation on the landscape, takes over. Lizards and small mammals find ground cover again. The Martu moderate a steady supply of different food types for themselves through this system, but they also create diverse types of habitat and forage for the other species on the landscape. The hunters thus provide seasonal larders for their prey, who thrive in balance with the environment. In five to ten years, the spinifex can "carry a fire," and it is burned again. If it is not burned now, the spinifex dies, dries out, and becomes susceptible to big, destructive conflagration. As Bird writes, "When Martu say that hunting and gathering is necessary for the perpetuation of life, it's not only an esoteric statement . . . it is based in a sound ecological theory of trophic interactions."

The Beauty-Death Transaction

Still, there is something fundamentally difficult to accept about predation and fire, and it has to do with pain, blood, and destruction. Physically, those are hard enough to bear, but it is perhaps even harder for us to grasp the terrible impersonality of death. Joseph Campbell was all about being as specific a person as one can be in order to finally let go of all ego identification in as fulfilled a manner as possible. He was constantly describing the paradox of expressing individuality while participating on behalf of the common good. This is the job description of the citizen scientist. But sometimes the details one documents are hard to stomach.

In spring 2013 I walked around Audubon Canyon Ranch in Marin County, where two species of egrets were nesting. Some of these prehistoric white birds were already sitting on eggs due to hatch around Mother's Day. On my way up the short loop to the main observing spot, I stopped at a lookout over Bolinas Lagoon. The low tide made it a mudflat. The docent offered me a telescope and a look at an unusual duck. "But what's that there?" I said. It looked like two deer were mucking about at the shore, but something was not quite right. I picked up a pair of binoculars. A small female deer was wobbling around with a bloody haunch. Her companion was not another deer.

"Life is, in its very essence and character, a terrible mystery," Joseph Campbell said, "this whole business of living by killing and eating."[20] It took me a few minutes to register what I was seeing. The doe—there were no antlers or antler buds on the animal's head—was clearly injured, but not otherwise discomfited. She took a few staggering steps. And a fluffy coyote lifted its head and lunged straight at her bloody wound.

Eating. The deer seemed tolerant. The coyote prodded its prey, taking a moment to keep the doe within its control but then, with seeming nonchalance, the coyote stepped away to chew more thoroughly. I was transfixed and repulsed. This was the trophic cascade in action. No wonder we can't stand to really think about it. "It's so weird that I'm seeing this right here on the lagoon!" I said out loud to the docent, who was staring with an open mouth.

In the not-too-far distance, a big sandbar was ringed with sunning black ellipses—seals—punctuated regularly by little black commas—their young. The way they curved and the way the sandbar curved was just like the way the reef and the cliffs curved. The scene was perfect.

"Formerly you had a dreamtime paradise," Campbell said. "No time, no birth, no death—no life. The serpent, who dies and is resurrected, shedding its skin and renewing its life . . . is the primary god, actually, in the Garden of Eden," he said, explaining the Sumerian origins of the story in Genesis. "When you look at the beauty of nature, and you see the birds picking around—they're eating things. . . . The serpent is a traveling alimentary canal, that's about all it is. And it gives you that primary sense of shock, of life in its most primal quality. There is no arguing with that animal at all. Life lives by killing and eating itself, casting off death and being reborn, like the moon."[21]

I continued on the trail to watch the egrets. There were plenty of them, snowy egrets and great egrets, doing their plumage showing, rearranging the twigs in their nests. Another docent said, "The males and the females spend equal time sitting on the eggs. They share all the household duties." The egrets were so pure white, their feathers so soft and luminous against the dark redwood branches, it cooled my mind to watch them. Until I remembered that when the chicks hatch, Mom and Dad will look on while the two biggest siblings kick the smallest hatchling out of the nest.

I returned to check on the doe-coyote drama below. The doe was dead, and a turkey vulture was having at it. "A man got out of a car to see what was going on," the docent told me. "The coyote scooted away." The docent then went on to tell me the deer had gotten stuck out in the mudflats earlier in the day and a sheriff and humane officer had lifted her out on a canvas tarp, depositing her safely on land, so they thought. He told me it was likely the coyote had chased her into the mud to begin with, but the officers hadn't known about that. Perversely they had helped the coyote kill the deer—indeed we are so often doing something other than what we think we are doing. The docent shook his head at nature red in tooth and claw, and remarked that even the benign-looking seals herd fish up onto the mud where they can't get away. And then I left Audubon Canyon Ranch to find … lunch.

Joseph Campbell's elucidation of the alimentary canal in the Garden of Eden may be one of his most significant connections between myth and science. Here in perhaps the central story of Western civilization lies a deep intuition about how nature works, what is really going on here. It is a lot about consumption—not just food we put in our mouths but the other natural resources we break down and convert into energy.

The energy transfer is day to day and also over longer time. My father has *passed*, I have mused to myself in the time since he died—I know the term refers to his soul now gone on to some other place, but it also describes a transfer of energy, those carbon molecules of his body now breaking apart and reconstituting as earth, water, and sky. One impact of his death is that it has bumped me closer to the category of "older generation." The word *generation* meaning both the making of energy and the embodying of it. We're all just carbon carriers with big burdens in our hearts.

Campbell also said that the universal story of the hero's journey got started when *Homo sapiens* became hunters. Agriculture was a collective undertaking, but hunting relied on an individual who had to go do something life-threatening. Hero myths helped hunters identify with a larger purpose for risking their lives, invoking power and meaning for the journey out with a spear. Many indigenous practices include a liturgical ceremony acknowledging a downed animal's connection with its forbears, also acknowledging the intertwined fates of human and animal ancestry, bound by the sacred act of killing. It's an intimate act but also, as with the doe and the coyote at Bolinas Lagoon, fundamentally impersonal.

Native American prayers over the killed animal acknowledged the generational transfer of energy, and an ecological balance. We market the hero's journey as self-determination, but what a pale project compared to the comprehension of our place in the trophic whole. Are we the heroes? The large carnivores that we have done such a good job of reducing, grizzlies and wolves among them, are animals that leave their natal home, light out for a huge territory, find a mate, and establish a new base of operations. Animal movements were likely the impetus for *Homo sapiens*, who followed them, to create the first maps. Today, as we examine the data points of their activities, we understand that their movements across the earth impact deep down into biotic and abiotic functioning of said planet. The life history of a wolf or a mountain lion thus creates the blueprint of a cosmic hero's journey, extending past spatial and temporal boundaries to encompass one history.

CHAPTER FOUR

Dream Machine

I stood at my kitchen counter in San Francisco, working on my father's obituary. I was almost embarrassingly quick to sign up for the task. The owner-editors of the *East Hampton Star*, the weekly newspaper in which the obit would appear, are lifelong friends of my family. Any number of the Rattray family could write a very solid obituary of my father. "Let Helen know I'm doing it, okay?" I asked my mother barely twenty-four hours after his death. "Because she may very well already be on it."

I was on to myself and didn't bother rehearsing any bogus excuses for why I must write this; his obituary would be one occasion upon which my father would insist on the facts, regardless of whether "spoiled by an eyewitness," and I would see that the facts were straight. There were so many facets to his life, so many people with different experiences of him—it would be impossible for even Helen Rattray to capture him, for example, as he trucked through the world of advertising. She wouldn't necessarily know how important it would be to him to have his military service highlighted. He aligned his own consciousness with that of Ernest Hemingway, who in turn was acquiescing to the rhythms of nature, on his deathbed—this felt to me of the utmost to get across. I had about a thousand words.

None of which was the whole point. This was how I would control my experience of his death. I would contain my father in truthful words representing him in a way he would recognize and appreciate. I would have this last pass at his approval. I drafted two sentences. "Edward Hannibal, a successful novelist and advertising executive, died peacefully at Southampton Hospital after a short battle with lung cancer. He was seventy-eight." I was writing on a sleek word processor, but the black letters appearing at my comparatively light-as-a-feather touch found origination in those hunted and pecked with index fingers on his upright Royal typewriter in the attic of our old house. My concentration in channeling the details of his life took its cue from his at that desk. My focus picked up the baton from his focus and I am still carrying it. On another level, I sought some kind of parity, or maybe plain old control. He created me. Now I would re-create him.

I recalled the travails of a working-class son, the first in his family to graduate from college, which he gave narrative life in his 1970 novel *Chocolate Days, Popsicle Weeks*. The title refers to the factory-line schedule at an ice cream manufacturer where he worked to put himself through college. At a party Woody Allen told my father he liked the book but the title made him nauseous, which made my father roar with laughter. The book was a *New York Times* bestseller, and he was awarded Houghton Mifflin's literary fellowship. Then there was the spell of writing full-time in East Hampton, the subsequent, less-successful books, and the return to Madison Avenue. He was one of the first customers of the Hampton Jitney, climbing aboard a single van that left East Hampton on Monday morning and returned from New York City on Friday afternoon.

Encapsulating his advertising career, I was struck with the originality of his ideas, the way they popped, the merging of image, word, and message that great advertising delivers. There was the third leg he gave O. J. Simpson in a print ad for Dingo boots. His favorite accounts were fragrance and cosmetics. This was where he got to ponder beauty, which hardly needs promoting; he cast the impossibly beautiful Veronica Hamel in a Revlon ad well before her fame on *Hill Street Blues*.

The then director of the Mitsubishi automobile account, David Stickles, perhaps overattributed to my father the invention of a whole new way of doing advertising the "un-advertising" way. This would be the precursor to any spots you might be tempted to watch on YouTube—little films that seem incidentally to feature something for you to buy. To associate a new sedan with the promise of family life to a young buyer, Stickles told me, "Ed parked the car out in front of a very pretty New England church, tied some tin cans to the back bumper, and put a Just Married sign on the back. The camera just panned around the car, always holding the church in the idyllic scene." The voice-over intoned Corinthians ("When I was a child . . ."). By the time the wedding music finished and the biblical phrase was completed (". . . I put away childish things"), the client was awestruck. "As they should have been," Stickles told me, "because Ed did something no one else had ever done." I was taken aback at remembering the sheer blasphemy of this wonderfully successful ad. The altar boy of his youth had been summarily dismissed. As Joseph Campbell saw our time, I saw that here, "every last vestige of the ancient human heritage of ritual, morality, and art is in full decay." A 1991 Smithsonian World documentary featured my father prominently. Its title: *Selling the Dream*. Inwardly and outwardly he balked, but he was very good at full decay.

Founding Citizen

There are people who don't like the term *citizen science* because it would seem to exclude people who are not currently citizens of the United States. Every time I hear this argument I find a way to eventually tug on the sleeve of the doubter and ask him to please consider reclaiming the word *citizen* from its current immigration-policy usage, and plant it back where it belongs, in the sense the mid-twentieth-century author and ecologist Aldo Leopold invoked. He advocated we reenvision our place in nature. The individual is not a dominator but a "plain member and citizen" of the land.[1] Leopold's association finds resonance in the roots of the word *citizen* with the idea of dweller—initially a citizen was a city-dweller, then a country-dweller. By the late sixteenth century the Greek term *kosmopolitês* indicated a citizen of the world, which is essentially the situation of each of us today. To concede the basic point, *citizen* has been used as an expressly political category. This isn't a reason to jettison the term—it is a reason to more heartily embrace it. Citizen science is not just an expression of an achieved democracy; it is a vehicle for creating more democratic transactions. This dimension of citizen science is not lost on the institution that officially supports our democracy—the United States of America.

In September 2015, citizen science got a multilevel endorsement and boost from the federal government. Addressing a citizen science forum at the White House called "Open Science and Innovation: Of the People, By the People, For the People," the director of the Office of Science and Technology Policy, John Holdren, observed that "citizen science and crowdsourcing enable research at large geographic scales and over long periods of time in ways that professional scientists working alone" can't easily duplicate. He added that citizen science can "create a sense of connectivity, community, and ownership" of projects addressing societal needs and noted that participation in citizen science increases STEM learning—education in the areas of science, technology, engineering, and math. He directed that the practice be held to the same standards as traditional science, and that the data collected by federal projects be "transparent, open, and available to the public."

At the time of Holdren's speech, nearly two dozen federal agencies were already running more than eighty-five citizen science projects—there are likely many more now. In full-on sponsorship and encouragement of the concept, he announced that federal agencies would each designate a coordinator for citizen science and that the crowdsourced work of the agencies would be collected by the General Services Administration in a new database, visualized on a new Web site: citizenscience.gov. He also announced a toolkit to help agencies design, carry out, and manage citizen science projects. More than 125 federal employees from more than twenty-five agencies developed and tested the toolkit, making it "a lesson in collaboration in and of itself." In sum, at the highest levels of our government, citizen science is being seen as a tool and a platform for achieving our mutual goals as a nation.

Though she never fails to assert the collaborative, grassroots, iterative nature of how the Federal Community of Practice on Crowdsourcing and Citizen Science cohered, Lea Shanley is one of its key architects. As director of the Commons Lab at the Woodrow Wilson International Center for Scholars, Shanley told me, she and her colleagues worked with volunteer groups like the Humanitarian OpenStreetMap Team on relief efforts in the aftermath of Haiti's devastating 2010 earthquake. "People were curating social media and mapping the streams of tweets, communicating with each other and with relief workers about where people needed food, shelter, and general assistance," she told me. Talking about the origins of citizen science as we define it today, Shanley pointed out to me the fundamental "gift to humanity" that makes it possible: Linux open-source software, made ubiquitously available in 1991. The geospatial, or mapping, community quickly purposed Linux to visualize knowledge about who, what, when, and where, in the service of democratic goals. "The mapping and the biodiversity communities emerged at the same time, evolved in tandem, and are now intermingling."[2]

If you would like to get your patriotic juices flowing, contemplate the small group of representatives from government agencies focused on disaster response who started to get together and talk about what was going on. They noticed that the same opportunities for quick communication, situational awareness, collective problem-solving, and action could be applied in other directions—for example, in environmental monitoring, biomedical research, and telecommunications. They were concerned, however, about data quality, policy impacts, privacy, and liability issues. Meetings ensued, with a growing group attending on their own free time; people, Shanley averred, "who work extra hard." She mentioned working every weekend and into the evenings for four straight months to help organize the White House citizen science forum and to collaborate on drafting the Holdren memo. I asked her about the passion behind her commitment and she laughed tiredly. "Members of our community are very public service oriented and believe in participatory democracy and science," she said. "Open data." A core group of thirty to forty federal staff volunteered consistently, and more than 125 people helped build the government's citizen science toolkit. With a $1.8 million grant from the Alfred P. Sloan Foundation, the Wilson Center funded an array of nineteen projects to explore citizen science at the interface of government agencies. Shanley and the federal community also worked to integrate citizen science in the president's Strategy for American Innovation, among other strategic initiatives.[3] A projected annual Citizen Science Day, inaugurated on April 16, 2016, was organized with the significant assistance of Jennifer Shirk (from Cornell University) and Darlene Cavalier. Cavalier's foundational Web site is a resource for "science we can do together." (Check it out: scistarter.com.)

The Map and the Territory

Aristotle, Leonardo, and Darwin are trotted out as original "citizen scientists," and so they are, in the definition of one who explores the world of causality from the perspective of the amateur. Most of what people refer to today as citizen science has to do with figuring out patterns, like trying to discern what is going on with sea stars in the ocean. But citizen science has a huge other dimension. Here Thomas Jefferson is not quite our model, since we must make many modifications to his worldview, but a touchstone nevertheless. He brought a democratizing perspective to the landscape itself. He sought to figure out a way to build a nation based on using and not overusing natural resources, which he envisioned being distributed fairly.

Jefferson supported two of the most significant citizen scientists in American history, Meriwether Lewis and William Clark. The expedition of their Corps of Discovery takes its place in the historical lineup of quests for empire established by the Spanish, English, and French, but with a critical difference. Like other colonial expeditions this one set out to discover the unknown and to locate sources of revenue. Beaver were being methodically extirpated as the terrestrial fur trade inched its way west, for example, and Jefferson was expressly looking to find more beaver and easier ways to transport the pelts east. But the Corps of Discovery was also sent out to survey territory upon which to extend a United States that would be big enough and powerful enough to stand independent of the world's other sovereignties. It would be equal in strength (at least) with other nations, but built on entirely, well, revolutionary principles of fairness and equality. His main tool in all this was the map.

As *The Log from the Sea of Cortez* would several hundred years later endeavor to reconceive the expedition journal, Jefferson made his own distinctive mark on the genre. In 1781 he completed *Notes on the State of Virginia*, his only full-length published work. As Steinbeck and Ricketts do, in his own way Jefferson approaches his subject as a dual thing, both an inventory of a terrain and a disquisition on meaning. Literary critic Leo Marx writes, "It is in its way a minor classic, rising effortlessly from topographical fact to social analysis and utopian speculation."[4] Most of the book is an inventory of the natural resources of Virginia such that would make a US Geological Survey scientist proud.

Rearranging a sequence of questions perhaps competitively posed to him by François Barbé-Marbois, a French diplomat, Jefferson frames his book as an explanation. Thus, Query I: "An exact description of the limits and boundaries of the State of Virginia?" Lats and longs are given. Query II: "A notice of its rivers, rivulets, and how far they are navigable?" As we might point to a data visualization on Google Earth today, Jefferson responds that looking at a map of Virginia "will give a better idea of the geography of its rivers, than any description in writing." Still, he's thorough and provides a rundown. "The Ohio is the most beautiful river on earth," he declares, listing a variety of measurements, including its width at multiple points and lengths from its origin in Fort Pitt, Pennsylvania, to twenty-eight separate locations. He measures winter and summer tides. He counsels on maritime traffic through Virginia's waterways and adds a side note: "In case of a war with our neighbours, the Anglo-Americans or the Indians, the route to New-York becomes a frontier through almost its whole length, and all commerce through it ceases from that moment."

The contents of scientific inquiry, whether collected by amateurs or professionals, are eagerly purposed by those in power or seeking power to further territorial goals. Jefferson is often invoked as a kinder, gentler, or at least ambivalent imperialist, but he is not an exception to this general trend. Jefferson owned slaves and his decisions helped vastly reduce Native American agency. The sunny side of Jefferson's image today inspires what's called a "new agrarian" movement to empower small, self-sufficient farms across America as an alternative to the destructions of industrial agriculture. It's perfectly possible to farm and graze livestock in ways that not only do not degrade the environment but even help increase biodiversity, sequester more carbon in the soil, and support healthier waterways. It's just not usually done that way.

In responding to Query XIX, concerning Virginia's manufacturing and trade, Jefferson notes that in Europe the land is all either under cultivation or prevented from being cultivated, so an economic source for supporting most of its people is of necessity found in manufacturing. But in America, "we have an immensity of land," so no shortage, and he questions therefore why half the people on it should be pulled away to "exercise manufactures and handicraft arts for the other." In what Leo Marx calls the locus classicus of Jefferson's pastoral American ideal, Jefferson proclaims: "Those who labour in the earth are the chosen people of God, if ever he had a chosen people, whose breasts he has made his peculiar deposit for substantial and genuine virtue. It is the focus in which he keeps alive that sacred fire, which otherwise might escape from the face of the earth."

Then he says that there are no examples in history of corrupt "cultivators." To farm is to be good.

Selling the Garden

Jefferson envisions an ecologically balanced America, in which we produce what we need sustainably, and not in excess or to accumulate wealth and power. He also favors what today we call quality-of-life values over pure economics. Almost immediately upon publication of *Notes on the State of Virginia*, Jefferson retracted his proposal of an undeveloped America without manufacturing or industry, calling his idea a "theory only, and a theory which the servants of America are not at liberty to follow."[5] Why the quick turnaround? Marx suggests that Jefferson knew it would take legislation to restrain economic development, and this would bolster the power of the government. To give more power to government for the sake of what he's tacitly admitting is a dream would contradict the heart of the new nation's enterprise.

Jefferson was caught between a rock and a hard place. To spur economic development would mean corrupting the character of what William Empson called the "mythical cult-figure" of the pure farmer cultivating his piece of Eden.[6] To discourage manufacturing and trade would consign the new country to a status dependent on Europe, with the distinct possibility of being forcibly annexed by a stronger nation. Jefferson's thinking in general can be looked at as self-contradicting—here was a guy who wanted to stay at home at Monticello, but became the third president of the United States. Marx imputes to him the quality of "negative capability" John Keats ascribed to Shakespeare, the ability to move fluidly between "uncertainties" without needing to find false resolutions. According to Marx, Jefferson's mind "moved ceaselessly between real and imagined worlds."

Jefferson's idea for a cultivated rural continent expressly invoked the vision of America as a farmed Eden. Asserting that Jefferson was the "intellectual father" of the nation's advance to the Pacific, the scholar Henry Nash Smith commented in his 1950 book *Virgin Land* that the Lewis and Clark expedition was important mostly on the level of drama and imagination. It enacted the Edenic myth from which would flow a righteous future. People moving west "plowed the virgin land and put in crops, and the great Interior Valley was transformed into a garden: for the imagination, the Garden of the World." The Garden of the World was not the invention of Thomas Jefferson alone. Before him, Benjamin Franklin basically marketed the vision to the British. Seeking land grants from the Crown, he proposed that settlement of a vast territory of nearly unimaginable potential would be a wonderful thing for them: "What an Accession of Power to the British Empire by Sea as well as Land!"[7] Franklin wanted the Brits to see American farmers as a market for their goods. Full-on decay was full steam ahead.

Parched Garden

Farming happened here, to put it mildly. Apocalyptic-scenario seekers need look no further than many of the fields in California today—driving through the Central Valley can be an exercise in pondering the definition of *dry*, marveling at the dark gray ground and wondering where the rich black soil went and if it will ever return. The development of agriculture as the fateful turn our species took toward resource depletion, from wild species to fossil fuel reserves, is an ongoing question.[8] Certainly Big Ag is a major contributor to climate change, which has ratcheted up California's drought, and the land seems to be doing as the whales did to Charles Scammon's boats, striking back. But just as we are not likely to easily replace California Indian burning and other practices on the landscape, neither are we going to feed billions of people on small farms alone. We got problems. But maybe we can save some birds in the meantime.

Mark Reynolds works for the Nature Conservancy (TNC) and is largely responsible for one of the most innovative uses of citizen science yet devised. "Gretchen came home from a meeting," Reynolds told me, referring to his wife, Gretchen LeBuhn, who runs the Great Sunflower Project, which studies the decline in bee populations, and is a professor at San Francisco State University. "She said, 'Have you seen this?' I said, 'Yeah yeah, an eBird record-keeping, bird-watching thing.'" (eBird is the behemoth citizen science project of Cornell University's Lab of Ornithology, and a standard-bearer for the field.) But then he actually watched the visualization she was talking about.

TNC is a conservation organization that mostly buys or otherwise protects land from development. "I'd been trying to advocate for doing things differently somehow for migratory species," Reynolds told me. "We could wind up with big patches of expensive conserved habitat that aren't well aligned with animals that move. I'd been on that soapbox for a while, but I didn't have a good way to tell the story." LeBuhn showed Reynolds the visualization of a single species of songbird as it migrates across the continental United States. (You can watch lots of these visualizations here: ebird.org/content/ebird/occurrence.) A dark map of the lower forty-eight begins getting brighter at the bottom, and the brightness moves upward as the bird movements track with the date over the course of a year. The image is flashing, evanescent, something like a long glimpse of flapping feathers. Among wonky maps, it is beautiful. The data it is based on was collected entirely by citizen scientists documenting when and where they saw this particular species of bird and sending that information to the Cornell Lab of Ornithology.

"I said, 'I can use this to determine when species are going to be where,'" Reynolds told me. "I thought it would be a great way to build enthusiasm around the idea of protecting bird migrations. But I was still thinking along the model of buying a farm or an easement and restoring part of it." The Central Valley of California where Reynolds was working has been almost entirely converted to agriculture, industry, and other development. What was once about four million acres of wetlands and associated habitat has been reduced by 95 percent. Historic habitat is no longer. Migratory birds need water. They need places to find food and spend the winter. Reynolds was working with the idea of patching together a corridor of protected landscapes so the birds could make their way across it. "We were despairing a bit," Reynolds confessed. "We have this one hammer, which is buying land, and it makes everything look like a nail."

Reynolds showed the visualization to his boss, who was enthusiastic about using it to help pinpoint with greater precision where to buy ranches or farms. Reynolds happened to share the details of this with a new TNC hire, Eric Hallstein, an economist. "Have you thought about a reverse auction?" Hallstein asked. "Just renting this habitat, not buying it? Could you get farmers to keep their lands flooded for the birds for a few extra weeks?" Working this out on the proverbial back of a napkin, Reynolds and Hallstein invented what is now called BirdReturns. It significantly pushes forward the uses to which citizen science data can be used to conserve nature.

For three years now, TNC has held a reverse auction among rice farmers in the Central Valley. "Their margins are so narrow, if it doesn't take much doing to make more money, then it's worth their while," Reynolds explained. "It's a way to extract economic value for their land and do a good thing." Working with data analysts from Point Blue Conservation Science as well as Cornell, Reynolds starts with visualizations from satellite data that show where water is likely to be over the course of a year, based both on weather patterns and agricultural practices. Data from eBird shows where particular species are likely to be over the same time period. A third map is produced, predicting where birds and water are likely to overlap, and not, and when. Rice farmers are then enjoined to flood their fields at strategic locations and times to support the birds as they make their way across the landscape.

The maps are used to predict, but Reynolds also enjoins citizen scientists to corroborate by resurveying or ground-truthing the predictions. He's established a feedback loop whereby citizens participate in a project, are apprised of the results, test the results, and then do that again when the project parameters have been adjusted. Most citizen science projects have people collect data, and that's it. But to truly get involved with nature requires more than one-off counting. The benefits of this kind of sustained involvement include getting a closer look at the heroic migrations of shorebirds. They travel staggering distances and have done so for thousands of years, darning up the web of life from one end of the globe to the other.[9]

Surveying the Garden

In 1793 Jefferson obtained backing from the American Philosophical Society to send an explorer to the Pacific and back. An eighteen-year-old Meriwether Lewis volunteered to lead, but Jefferson instead chose a French botanist, André Michaux, for the job. The trip's précis and instructions prefigure those Lewis eventually did get to follow, including the directive "to find the shortest & most convenient route of communication between the US. & the Pacific ocean, within the temperate latitudes." This was most likely to be the Missouri River, so "as a fundamental object of the subscription, (not to be dispensed with)," the river should be explored, and because the landscape in question belonged to Spain, to get there Michaux should cross the Mississippi north of the Spanish garrison in St. Louis. He was from there to head west to the Missouri, cross over the mountains he would find there, and from the Columbia River head to the Pacific. Jefferson instructed Michaux to "take notice of the country you pass through, it's general face, soil, river, mountains, it's productions animal, vegetable, & mineral," and to inventory latitudes as well as "the names, numbers, & dwellings of the inhabitants, and such particularities as you can learn of ... them." The journey began but was preempted. By the time Michaux had reached Kentucky, Jefferson discovered he was a secret agent of the French Republic, his mission to attack the Spanish holdings west of the Mississippi. Jefferson aborted the Michaux expedition.[10]

Ten years later, Jefferson gave Lewis the nod. According to the historian Stephen E. Ambrose, "Evidently he consulted no one, asked no one for advice, entertained no nominees or volunteers, other than Lewis. This was the most important and the most coveted command in the history of exploration of North America." And Jefferson was giving the job not to a "qualified scientist" but basically to a labile and eager student. Later Jefferson drew the portrait of the whole person rather than the narrowly defined specialist it was necessary to put in charge of such a quest: "It was impossible to find a character who to a compleat science in botany"—Lewis was tutored in this by Jefferson—"natural history, mineralogy & astronomy, joined the firmness of constitution & character, prudence, habits adapted to the woods, & a familiarity with the Indian manners & character, requisite for this undertaking. All the latter qualifications Capt. Lewis has."[11]

Ambrose remarks more than once in his book *Undaunted Courage* that the reports made by Lewis and Clark on the Corps of Discovery expedition were modeled on *Notes on the State of Virginia*: they were "part guidebook, part travelogue, part boosterlike promotion, part text to accompany the master map." While the journals provided fresh information about the flora, fauna, geology, terrain, and Native Americans to be found in the western reaches of North America, the star of the Lewis and Clark show was the map they made of their journey. It laid out an express plan for what would be called Manifest Destiny. It was an instruction diagram in how to make a strong nation out of a nascent one.

After the Gold Rush

As sea-to-shining-sea was in the process of getting official boundaries, Congress deployed a practice of portioning each state in equal measure, though California contrived to define itself, and the result is darn big. In *How the States Got Their Shapes*, Mark Stein points out that primarily because of its natural resources, "the United States needed California more than California needed the United States."[12] Just as California was becoming part of America, gold was discovered in the foothills of the Sierras. "California's eastern border is one of the few items from the Gold Rush that is still on the ground," explains Stein. "Its existence is evidence of how important it was to California to possess all of the gold-bearing mountains in the region." Congress lobbied to apportion the wealth differently but had to concede whatever California wanted. The Golden State had wealth not just buried in the earth but growing out of it in the form of redwood trees. Vast oceanic resources (sea otters and whales among them) sustained generations of international trade, and seemingly endless tracts of land on which to graze cattle and grow crops helped establish an economy that today is the eighth largest in the world. The history of all of the United States is predicated on economic purposing of natural resources, but California has an awful lot of natural resources in one place, and American possession of them came at a technological moment in which these could be transformed into money very quickly.

The gold rush almost overnight transformed a remote outpost, San Francisco, into the hottest destination on the continent. Sojourners from the East either had to take the long way down and around South America to get here, or else had to come across what was still largely the Wild West; the city was for a time burdened by the quantity of schooners abandoned in the bay by people docking and taking off into the hills. In the first pulse of what subsequently became periodic explosions of wealth, San Francisco was infused with money and its ardent seekers. As Stein said, "California was suddenly filled with a population, an economy, and a very high crime rate."

We think of gold as coming from the earth, and so it does, all the way back into prehistory, when the planet was molten. Gold arrived here by way of meteorites, and as the earth formed, the heavy element sunk to the core. As geologist Mary Hill writes, "Gold is in the sea as well as in most rocks."[13] Hot, fluid rock, or magma, coming up from the core of the earth, carries gold along with it. As lava pours out of volcanoes midocean, it cools, cracks, and reacts chemically with the salt in seawater. Gold and other metals in this hot batter are dissolved and spread out, then get concentrated like hunks of chocolate in a generous cookie. The gold that changed California forever was and is (there's still some left) mixed in especially with the granite and sediments that accumulated on the western side of North America in what geologists call the Smartville complex, named for a little town near Grass Valley.

The Smartville complex is an ancient immigrant, made in ocean depths elsewhere, finding its way to "future California," as Hill puts it, about 160 million years ago. The complex, or "belt," is twenty-five miles at its widest and runs from Auburn to Oroville, about fifty miles long.[14] According to two alternative scenarios, the Smartville complex either rode over two ocean trenches, called subduction zones, and destroyed them, or else in the process of diving into one of them had a big piece of itself sheared off, which is now sitting there like a rocky, sparkly cap. What is called the "mother lode," where most of the forty-niners' gold came from, can also be identified as the "Smartville suture," and Hill calls it "a piece of ocean crust, a very surprising batch of rocks to find in today's Sierra Nevada."

As the terrestrial landscape is itself the product of great oceanic forces, perhaps most especially asserted in mountain ranges like the Sierras, so is this most treasured extractive produce brought to you courtesy of H_2O. And wasn't James Marshall surprised when he found a flake of gold in the water while working for James Sutter, who was kind of irritated that his big plans for an agricultural empire would now get all overrun with a rush for gold. He tried to keep the discovery a secret but we all know how that is.

While mammalian natural resources lose their lives to serve human industry, gold's even more concentrated value lies in its ability to keep its essential profile even while undergoing a vast variety of shape-shifts. Gold stays gold. It doesn't oxidize and it doesn't corrode. It can be mixed with other metals to form alloys, but it retains its own chemical composition. Its ability to carry electrical current makes gold a favored element in electrical contacts, computers, and electronic and space equipment. Air Force One is coated in gold, because it deflects and confuses the signals of heat-seeking missiles.[15] From the depths of the earth comes this true thing of enduring value. Gold has an illustrious as well as lustrous history. For centuries, it was the basic unit of currency for civilizations including the ancient Greek. (It reflects some of the centrality of the beaver to the establishment of North American European populations that for a time a beaver pelt conveyed this same status and equaled money. Fur was gold for a time.)

So, we want gold. The race for gold in the 1850s was a hysterical compulsion, a take-no-ecological-or-human-prisoners incentive program that wreaked destruction at every turn. California waterways were summarily purposed. As Hill puts it, "Without water, mining can barely work." Just for starters, even the low-tech strategies of panning for gold, and using a "cradle" to coax it from the water, kicked up sediments that immediately affected salmon and other fish runs and destroyed streamside vegetation. Hydraulic mining, which was accomplished through dams, reservoirs, and other man-made adjustments to the geological and hydrological water cycles, impounded vast quantities of water and supplied it to miners along seven thousand miles of ditches.

Statehood established, a California state surveyor was named in 1853. Geologist-explorers brought European-based science to the West even as their work had expressly political purposes. Land was dispensed according to the way it was mapped. It's worth lingering on this point. The way we conceive of the landscape, who can use it and how, who "owns" it, what even it is actually composed of—are its valuable qualities natural beauty or commodities?—finds expression in the map. The map dispenses our attitudes toward the land and helps turn it into dollars and cents. Federal surveyors mapping California in the mid-nineteenth century worked at a triple intersection. They were pioneers of scientific discovery, the first Westerners to explore much of the terrain. They served political purposes, as the lines they drew determined name, ownership, and social and commercial uses of the land. And they were celebrators of an emblematic destiny, of a sublime attainment expressed in the geology itself. In *Pacific Visions: California Scientists and the Environment, 1850–1915*, Michael Smith says, "Scaling the mountain, charting the terrain: in the symbolic conquest of the West, nature itself seemed contained in the nets of longitude and latitude cast by the pathfinder."[16]

At this point most of the territory had been divided up by Spanish and Mexican land grants called ranchos. Now the Public Land Survey System originally proposed by Thomas Jefferson in 1785—following his dream of a yeoman farmer nation—was brought to bear on redrawing those lines. In the Public Land Survey System are discerned the guidelines he followed to write *Notes on the State of Virginia*. Property lines were to be based on topography and local markers and to follow a rectangular "system of survey." Townships were determined to be six miles square, each township to contain thirty-six sections one mile square.

A far cry from the precision of latitude and longitude as brought to us by satellite geolocation today, this slicing and dicing of the landscape was determined the old-fashioned way, by hip chain. Two "chainmen," using a specified sixty-six-foot measuring chain made up of one hundred links (Gunter's chain, named for the sixteenth-century astronomer who came up with it), literally swung the chain's length out, moving fast and counting up eighty chain lengths per mile. The surveying system was fundamental to identifying railroad right-of-ways. Subsequent placement of roads and suburban housing developments frequently branch out from these original demarcations. The Homestead Act of 1862 allocated 160 acres, or a quarter section of land, to settlers who would "use" the land, and later amend-ments made that number 640 acres, or a full section, when the land was not as putatively desirable for farming. Western ranch-ing got established on the 640-acre plan. And the survey's basic arithmetic is the premise for the "forty acres and a mule" prom-ised to freed blacks after the American Civil War.

How neat, how tidy, how susceptible to (rampant) fraud. Also, how unconscious or willfully blind about the blended, perme-able boundaries of Native California tribal territories. In their essay "A Brief History of People and Policy in the West," Thomas E. Sheridan and Nathan F. Sayre write that "Native American claims to the continent were largely ignored, justified in part by the belief that the land belonged to those who 'improved' it under the civil rights of private ownership. This was in contrast to those who had only natural rights, as defined by John Win-throp, a founder of the Massachusetts Bay Colony, describing a time 'when men held the earth in common, every man sowing and feeding where he pleased.'"[17] The evidence of Native land cultivation was totally invisible to the new settlers, who wanted any Indians who survived their aggressions to get with the pro-gram and become yeoman farmers on small chunks of what was now turned into private property. "Tribes lost an estimated 91 million acres of reservation lands between 1887 and 1934, when the Indian Reorganization Act terminated the allotment program and gave tribes a little more control over their natural resources," write Sheridan and Sayre. "They've been struggling to assert their sovereignty ever since."

Thomas Jefferson would not have liked what happened next. While the Homestead Act did encourage small farms, little or no scrutiny was brought to bear on the more prevalent trend for wealthy or just clever individuals, railroads, and other corporate interests to accumulate vast holdings of freshly available land.

The federal government had its eyes on the prize, which was to get the land "settled." As the process jostled forward, more "acts" created venues to make it easy for (white, male) citizens to claim land for small fees. Land with sequoias and redwoods on it went up for sale like the rest of it.

The very quick purposing of territory that followed soon upon completion of Jefferson's Corps of Discovery expedition is of course a seminal marker in American history. The time period is often called the "winning" of the American West. Jefferson used Lewis and Clark's map to help create one continental citizenry, but the mostly European-descended whites who braved this new world were in fact usurping a citizenry that had "settled" the land in ways they didn't recognize or value. Their methods were frequently unfair, unwise, and evil—given the genocide trained on the California Native peoples—and depended on obfuscating territorial claims. The gold rush, the laying of the railroad, the massive logging of the redwood forests—all of these activities that "created" America, and California in particular, also destroyed ecological functioning and health. The double narrative was in full play and its effects are still playing out. The map, however, is now poised to collapse the distance between creation of territorial claims and destruction of natural resources, by way of technologies that are freely available to citizens of the world. Here's an information revolution of which Jefferson would be most appreciative. Today's digital maps tell narratives like the one he produced in *Notes on Virginia*, combining the lay of the land with its history and its potential for supporting human enterprise.

CHAPTER FIVE

Green Thumb in a Dark Eden

The newer Silicon Valley campuses of the digital dynasties playfully signify that theirs is a new kind of big business. Facebook heralds itself on the back side of the original Intel Corporation sign that once announced its own brave new world; this is Facebook's modest gesture as if to say, "We're king now but so were they, not long ago." Google Earth Outreach is located down the road from the vaunted Googleplex mother ship in one of a honeycomb of roundish buildings. This is the place from which comes the power both to apprehend the celestial sphere and to find one's exact location within it. You can fly anywhere (virtually) on Google Earth. You can circumnavigate the globe, you can visit outer space, you can dive to ocean bottoms. We are all Sir Francis Drake and Captain James Cook now. My more immediate challenge the first day I visited Google Earth Outreach was the location of the front door to the correct building. A gigantic red teardrop like the little one on my phone marking my destination told me I had found it. The sense of having entered a screen was amplified by the street-view car parked out front, its sci-fi proboscis pointed nowhere at the moment.

Having just seen Rebecca Moore's cameo appearance in James Cameron's documentary TV series *Years of Living Dangerously*, I half expected to find Harrison Ford in her office. Her flaxen hair shining and swinging, she was sitting at a regular desk, albeit with a big screen behind her, upon which the virtual world turned. There was a "Wrong Island" map on another wall—this is a little geospatial humor. It shows Long Island flipped the other way, with the alligator mouth of the North and South Forks pointed left instead of right. Oh no, what happened to Brooklyn?

Moore conceived of and developed Google Earth Outreach, which basically brings the information and mapping power of Google to social and environmental justice projects all over the world. Since 2007, her team has nurtured a long list of practical, accessible, mind-blowing strategies for addressing some of the world's most intractable situations. In her office she gave me a tour of some of the projects she's the most proud of: working to help identify and clear land mines all over the world, thus allowing displaced people to return home without fear of being blown up; working with Jane Goodall to monitor chimpanzees and help conserve their habitat; using satellite imagery, data, and multimedia to track and show the world the systematic destruction of more than 3,300 villages in Darfur. To this effort then president George W. Bush responded: "Nobody can deny this genocide now. You can see it on the Internet."

A substantial number of Google Earth Outreach projects are focused on the issues of indigenous peoples around the world. Moore's work with tribes was initiated when Chief Almir of the Brazilian Surui people appealed to her; he wanted help in literally putting his tribe on the map. Together Moore's team and the Surui have transformed what was a purely tragic unfolding of illegal logging and deforestation into a success story. Armed with smartphones, tribal members record illegal activities when and where they are happening. Enforcement response is orders of magnitude faster than previously. The Surui have since contracted with Natura Cosméticos to sell carbon offsets associated with the saving of their rain forests. "We started out with the idea that we would help local people map their cultural resources," Moore told me. "But we realized that if their land is threatened, their culture is threatened."

Moore and her team "lean in" on several projects that use their capacity, including Global Fishing Watch, in which fishing boats all over the world can be tracked in real time, thus aiding efforts to crack down on illegal take (globalfishingwatch.org). Jane Goodall's Roots and Shoots initiative, which aims to connect youth to environmental issues, among many other things, provides a template for starting co-created community projects anywhere (rootsandshoots.org/streetview).

To support a larger number of projects, Moore puts on a weeklong conference each year called Geo for Good. In October 2015, I started out in the back of the room at Geo for Good. I marveled at the rows upon rows of people from all over the world. Every continent but Antarctica was represented. The daily schedule included presentations from participants about their projects, and then tutorial sessions at multiple levels, from the basic, such as simple mapmaking (these are the ones I attended), to the more complex, such as using Fusion Tables (don't ask) and strategies for those who know how to do their own Keyhole Markup Language coding (I'm going for another cup of coffee now). On Thursday came "Tools for Storytelling." A featured project was the Nepalese platform StoryCycle, which

Google helped produce. Now you can take a virtual tour of Mount Everest (storycycle.com/everest). As StoryCycle founder Saurav Dhakal told me, he wanted to find and tell the stories of the indigenous Sherpas who lead the expeditions. "They have a life beyond that role," he said. "And there are small towns and villages all over Nepal with hidden and forgotten stories." In the *Nepali Times*, Dhakal wrote that he wanted to tell these stories, but not "in the old Ramayana way. We wanted to show the stories through pictures, videos, graphics and maps. And the stories through the people not through a Pundit."[1]

For those of us rank beginners, tutorial materials were pre-populated with photographs, coordinates (latitude and longitude), and textual descriptions to aid us in assembling a map of the explorer Ernest Shackleton's life and journeys, especially to the Antarctic. Of course, here the expedition met the story, met the Google map. I did it! I was aided by the kindly Brian Thom, an anthropologist from the University of Victoria in Canada, who walked me through some basic steps at which he is adept. In his work, Thom inserts multiple layers of information about indigenous people into maps of British Columbia's west coast (uvic.ca/socialsciences/ethnographicmapping). He enjoins teenagers of tribal descent to interview their elders, and their voice recordings and some video are embedded straight into the maps (so cool!). The landscape thus comes historically alive and Chronos meets kairos at the intersection of latitude, longitude, and time. The result is not just of cultural interest, however. As resource-extraction companies come calling to these communities, information supporting tribal claims on the land comes into play. As Thom put it, this kind of mapping makes "the difference between consultation and consent." Plain evidence "for aboriginal title will be represented in sophisticated maps crafted from traditional knowledge and indigenous experience of the land."

The Blue Line

Like Julia Parrish of COASST, Rebecca Moore is honored as a Champion of Change by the White House. Hers is a hero's journey with clear archetypal markings. As the schema articulated by Joseph Campbell in *The Hero with a Thousand Faces* has it, instigated by an internal crisis, Moore departed from a perfectly solid path by which she was conventionally achieving and ventured into the unknown. Following her instincts and her interests, her bliss, Moore underwent a protracted challenge. "Seizing the sword" for Moore meant harnessing and purposing computing power in a way no one had quite done before her. The elixir she brought back to her people (us) not only provides balm, empowerment, and justice in her case; it provides a tool by which the rest of us can also become heroic. Her story additionally has uncanny soundings in American history that Thomas Jefferson would not perhaps have ever anticipated, but would appreciate.

"I grew up on Long Island and in the Adirondacks," Moore told me, "marinated in nature. I wandered in the woods with my cousins and brothers and sister." At the knee of her grandfather, Frank Charles Moore, she learned the Adirondacks had almost disappeared owing to logging and extraction in the 1800s. "We needed to draw the blue line around it and keep it 'forever wild,'" Moore explained. She was referring to what was then Article 7 and is now Article 14 of the New York State Constitution, which in 1902 established a forest preserve in the Adirondacks, delineated in blue ink on the state map according to county and town boundaries using survey tracts, streams, and railway right-of-ways. The Moore family's summer retreat is located within this "blue line." "We had an old boathouse and applied to the park authority to renovate it," Moore told me. "But they said no. They didn't want any more development. And you know, that's fair.

I realized that the very qualities of the place we enjoy so much are protected by these restrictions, so we should abide by them gratefully."

"There's a long tradition of public service in my family," Moore said, putting it mildly. Her "nine-times-great-grandfather" Rip Van Dam was born around 1660 in New York and, in the absence caused by the delayed arrival of William Cosby from England to govern the province, acted in his place from 1731 to 1732. "Cosby came over and demanded half Van Dam's salary," Moore said. "Van Dam is described as strong, able, and none too bright, but he said, 'No—where were you, I've been governing.'" An enraged Cosby sued Van Dam, who contributed unsigned, scathing critiques of Cosby to the *New York Weekly*, a broadside he helped found. Cosby turned around and sued John Peter Zenger, the publisher of the *Weekly*, for libel—in those days, the crime of printing information that opposed the government. The Zenger trial of 1735 is today a landmark in the unfolding of freedom of the press as codified eventually in the First Amendment. Zenger's advocate in the case, Andrew Hamilton, declared it "not the cause of one poor printer" but "the cause of liberty," and won. "It was a big jurisprudence case," Moore said, "establishing that if a defamation is true it is not libel. Truth is a defense against charges of libel."

A subsequent ancestor was Samuel J. Tilden,[2] twenty-fifth governor of New York, whose crusade against corruption in government ultimately led to his takedown of William M. "Boss" Tweed, described by one commentator as "Grand Sachem of the most powerful political machine in the nation."[3] Running for president in 1876 against Rutherford B. Hayes, Tilden won the popular vote and was narrowly ahead in the electoral vote, save for twenty votes that remained in dispute. These votes were from South Carolina, Louisiana, and Florida.

"There was a judicial group appointed to see where the Florida votes should go," Moore told me. "Tilden needed just one to get elected—they awarded all the disputed votes to Hayes. People were telling my great-great-uncle, 'Put your foot down,' but he said, 'It wouldn't be good for the country.'" (Al Gore, who famously had a parallel experience in his 2000 presidential run based on disputed votes in Florida, and responded in basically the same way, makes a surprise appearance as a Rebecca Moore booster some 140 years later.) Tilden retired a bachelor, and when the dust from the working-out of his estate settled, it became the foundation of the New York Public Library.

The defending-what's-right gene stayed strong in the Moore family lineage. Moore's father, Earle K. Moore, was an epic advocate for equity and openness in one of the defining challenges of his own era, the civil rights movement, and on the communications platform of his day, television. "When it came to fighting the logging," Rebecca Moore said, flashing forward to the battle she waged on a Santa Cruz mountaintop, "I channeled him."

"It was the height of the sixties in Jackson, Mississippi," Moore recounted, herself a little girl at the time. Southern TV stations were in the habit of reporting on protest and unrest usually without any black point of view.[4] WLBT in Jackson attracted attention for its egregious racial bias when the station refused to sell time to Reverend R. L. T. Smith, the first black person to run for Congress since the Reconstruction. Smith got an assist from none other than Eleanor Roosevelt, who wrote about him in her weekly column. Largely because of Roosevelt's intervention, Smith was sold a half hour of airtime.

The station kept at it, however, while racial conflict continued to rage, cohering around the impending desegregation of the University of Mississippi. Black activist Medgar Evers asked for a federal investigation into the station's coverage of the explosive issue. The station was evidently entirely in favor of keeping things segregated and yielded not a minute of airtime to anyone saying otherwise. Evers was assassinated. "It was a huge outrage," Moore recalled. "Evers' brother Charles and the Reverend Everett Parker of the United Church of Christ brought suit against the station," represented by Earle Moore.

The case became about the right of the public to have a say in the use of broadcast licenses that theoretically were distributed for the public good. "My dad came up with a novel legal argument," Moore explained. "He contended that the number of people who live within the transmission radius of the station and who have a reasonable expectation of viewing good programming have in aggregate invested more in their television sets than the station has invested in itself." In other words, the public has an ownership stake in the station, since without them it wouldn't have the broadcast license. Justice Warren Burger ruled in favor of the United Church of Christ. The FCC appealed—just as Burger ascended to the head of the Supreme Court. Upon its second appeal, the station then lost its license, and, as Moore put it, "the citizens took the station over! I grew up with this. My dad and I were very close."

"I knew I was supposed to do something that mattered," Moore told me. "I didn't know what. Nobody said it to me, but it was always there." Moore went to Brown University and majored in computer science.

"There was such a thing?" I asked her, remembering the era of punch cards and giant humming machines (I tried to make a computer write poetry and barely passed that class).

"I fell in love with it," she said. "Fortran, Pascal, C—I love logic and the beauty and clarity of proving a theorem, the sense of reaching an essential truth. But I didn't want to be an academic mathematician." Moore loved the natural sciences, she loved

English literature, and she initially declared a biology major. "When I stumbled on computer science, I thought, this is it. It combines the beauty of logical thought and problem-solving but you can actually build something in the world with it. It was like a drug for us; we'd be in the computer center through the night—it's more creative than people understand. It's a beautiful combination of creative and analytic."

Moore expressed her love for nature and the outdoors as a mountaineer. "One of my heroes is Arlene Blum, who led the first women's climb up Annapurna." Moore has climbed the Himalayas, Yosemite, to the summit of Denali. "That was my wild and crazy youth. I learned a lot through that, about leadership and fear and guiding yourself and other people through treacherous terra incognita. I've been pulled out of crevasses and I've pulled people out of crevasses. It gives you some perspective when you're in Silicon Valley and people are freaking out over a deadline."

Moore eventually became a software developer at Hewlett-Packard. "We built a unified messaging system that delivered text and voice mails," she said. "We got patents and were successful in products and in the market. But we were not making a social difference." Moore's nagging conscience came front and center when her father and her brother, the artist Frank Moore, died back-to-back between late 2001 and early 2002. "I was floored," she said. "I quit."

In his 1993 painting *Yosemite*, Frank Moore depicted the national park in a state of unraveling. Looming giant sequoia trees frame the picture. They have awakened in humanlike expressions of bewilderment perhaps because their green canopies have turned into spiraling holograms of snakes. Eden the beginning point has become hell the ending point. But heck, we're on vacation. Smoke signals forming infinity loops loft toward the horizon. One smoke cloud contains the image of Mickey Mouse and another the insignia of the Playboy empire—commercially purposed animal forms. Moore created object-studded frames for much of his work, and this one is decorated with pinecones; the title of the painting is constructed from wood pieces as if made in camp. The tango of innocence and experience takes a low dip. An important chronicler of the AIDS epidemic, Frank Moore helped the Visual AIDS Artists Caucus launch the red ribbon in support of a cure. He died of AIDS at age forty-eight, having lived with an HIV infection for fifteen years.

To lose a father when he is seventy-nine is a loss but in the generational order of things. To lose a beloved brother in the prime of his life and at the height of considerable artistic powers is quite another. "I had a dark moment," Rebecca Moore told me. In describing her love of climbing she had told me, "I love the long-distance view, I love exposure." *Exposure* is a technical term for the empty space below a climber. So with twin pillars of her life suddenly gone, Moore's exposure, and the imperative to keep climbing, intensified. Fascinatingly, she would purpose the biggest view possible, another way of describing exposure, in what would be both an emotional and a practical response to her personal challenge. She would share exposure with the rest of us, on a map.

Moore went back to school in bioinformatics. "It was an exciting time in the field. The human genome was being sequenced. There was this promise of personalized medicine based on interesting computer science algorithms." Moore went to the University of California, Santa Cruz, where the human genome data is stored. At the same time, a local and seemingly mundane occurrence led her down a different path. Moore lives in the Santa Cruz Mountains, which are fairly remote and rural despite being a half-hour drive from Silicon Valley centers of the universe. One of her neighbors had made a 911 call for a medical emergency, and the ambulance took several hours to find the right house. "They got lost," Moore said. "I found myself curious—what kinds of maps did the emergency people have?" Moore's small community is arrayed on private roads home-owners maintain themselves, and people have their own water wells. She began to cogitate on the inevitability of earthquakes here. "We will be the last to get emergency response here," she reasoned, "so how are we going to take care of each other?" She realized she didn't even know exactly who lived exactly where or how to find them.

"The fire department was using a map someone had drawn in pencil in 1983," she said. "It was completely wrong, missing three-quarters of the houses. I didn't want to be confrontational or anything but I said, 'Uh, do you ever have the opportunity to update these maps?'" The fire department said they had the authority to do so but not the time.

Moore gave herself a "crash course in the art of digital mapping. Someone turned me on to this thing called Keyhole. It cost eight hundred dollars. I asked for a trial use for six months as a community person and they gave it to me."

Keyhole Inc. was developed to make satellite and aerial imagery available for commercial use over the Internet. Its clever name references a military satellite reconnaissance project called Keyhole, as if to make the point that with its technology, there are no secrets anymore.

"I started collecting all the satellite imagery of the houses, our emergency stations, all the built-environment landmarks," Moore said. "I made a big map, six feet by three feet. It's pretty great. I still have a copy of it on my wall at home." Moore brought the map to a community center meeting on emergency response. "People went nuts; they loved it!"

"I'm in school, doing bioinformatics, but I'm actually getting excited about the map. There are old logging roads and deer paths around us—we each have about five to ten acres. I thought, what if we created a community trail map out of all of them? People were really into it." One holdout complained that he had moved to their mountaintop to get away from people, and didn't necessarily want to be more easily found by them. "I said, 'Don't you like to hike? I have an idea of where we could put a hike on your land where you would never see another soul.'" Mr. Reclusive was "a computer-geeky guy who liked the idea of having a fully redundant trail map, so he signed on."

Just at this time, the Midpeninsula Regional Open Space District (a Bay Area greenbelt system) acquired about 1,100 acres of historic forest called Bear Creek Redwoods. Here Captain John C. Frémont had led the famed 1846 Bear Flag Revolt, a brief flare-up presaging the Mexican-American War that eventually resulted in the addition of California to the United States.

"They were doing a master planning process and wanted public input—should there be trails, historic facilities? I realized there was no public open space on the ridge where I live. It was all settled in the 1800s and it's private. But looking at the maps I saw Bear Creek came all the way up to the summit. I said, 'YES!' We could provide ridgetop access by coming down into Bear Creek from the top."

The Midpeninsula staff demurred that the land was too steep. Moore started "playing with Keyhole (which would become Google Earth) and examining the landscape." The challenge was to get people to accurately understand the terrain, and also to envision what a ridgetop trail would provide—which Moore suspected was exposure. "I wondered if you developed a trail up there if you could see all the way to Monterey Bay," she said. She hiked to the summit and took some photographs. Then she overlaid the actual view captured by the photographs onto a simulated terrain—this is essentially what Google Earth does now. Presenting to some Midpeninsula staff she said, "Look, we can take people to see this land virtually, to experience it! And you can use Google Earth yourselves to better grapple with how to manage the land. It's free." Midpeninsula was paying expensive consultants who were laying out flat paper maps of the area.

"At this point I was much more excited about this than about bioinformatics. I thought, there's something here. The satellite imagery is so powerful. The fact that you can overlay your own information, like the view from that summit, to visualize where the community wanted to go with its land, how to use and steward the resource—I realized, this is democratizing the map."

Moore founded the Mountain Resource Group, introducing it as a "community geoportal" for the Santa Cruz Mountains, envisioning an Internet geographic information system (GIS) "which aggregates and presents geospatial information important to local residents in this semirural region." Thanks to the ubiquity of the smartphone, most of us are by now well-acquainted with GIS, as it helps us navigate our way through quotidian terra incognita like restaurants we haven't been to before. But as Moore articulates in her founding document, GIS can cohere information "related to management of local natural resources and community assets, such as redwood forest and watershed."

Moore began to envision community geoportals to "unlock the power of geospatial information where it may be most needed—for everyday people in their everyday lives." She spent more and more time going deeper into the complexities of Keyhole, and started to report bugs she uncovered to the company. "I did this in a very technical way," she confessed, "like, 'I think this is a race condition with your server,' that kind of thing." She offered suggestions about fixing them. One day she got an email from the company that said, "Who are you and what are you doing?" Telling me about it, Moore laughed. She hadn't known what she was doing quite yet, but as Joseph Campbell points out, when you follow your bliss, the universe seems magically to help you. In Moore's case, that universe would be Google itself.

Say What?

Suffice it to say, when an ecologically aware Northern Californian gets a letter in the mail that says, "We're going to do a little bit of redwood logging in your backyard," the response is not likely to be a shrug. Yet one almost feels sorry for the San Jose Water Company. Just at the moment they decided to drop a "notice of intent to harvest timber" on the doorsteps of various Santa Cruz mountaintop residents in 2005, Rebecca Moore had refined her ideas about how to democratize mapping and why doing so is critically important. Keyhole, the software she had used to make her community map, had been newly acquired by Google, and Keyhole personnel had become acquainted with her through her technical interactions with their product. Based on her interest and her informed knowledge of the platform, Moore was invited to give a "tech talk" at Google, a regular opportunity by which Google employees get to cross-pollinate with other experts in their fields.

"This was just before Google Earth was launched," Moore told me. "I compared three different technical mapping platforms, and my conclusion was that Keyhole had the most potential to change the world if they just did these eight things." Moore begins her presentation with the "well-known" acknowledgment that the digital revolution is upending business as usual, and bringing information as power to "every nook and cranny of human enterprise." Her talk in general is endearingly peppered with the word *citizen*, as she enumerates the various ways citizens now have a direct line to "authoritative content": citizen journalists (bloggers), citizen content experts (Wikipedia), citizen broadcasters (podcasts), and citizen publishers (desktop publishing and PDFs). She articulates how geospatial content is now easily accessible and relatively easy to integrate. For example, land records and parcel data—the very substance of the land surveys—can be imported directly onto a map. This information can be combined with personal information such as "my training route," or with locations, say, from crime Web sites about what has gone on where. Moore points out that the budget- and staff-limited government increasingly relies on "a trained and educated citizenry" to help with public safety and disaster preparedness.[5]

In this exploratory document, Moore does not quite cohere the implications of the variety of points she makes. Her profound observation that "global warming and climate change, energy and transportation policy, land-use and natural resource policy" are all "geospatially related issues, both local and global," doesn't take the central place in the document it perhaps would if she wrote it today. She notes that "single-issue geospatially oriented sites are cropping up all over," and she lists among other projects "salmon-habitat monitoring." The fact that our current predicament begs for integrating multiple one-off projects into a map so that process, pattern, consequence, and choices about them can be discerned is not yet articulated here. But as she notes, quoting from an influential 1995 paper by geographer Eric Sheppard, "The development of GIS, or any other technology, is a social process."[6]

Moore pitched Google her concept that the company think beyond geospatially oriented e-commerce to "offer something arguably more profound: a tool for citizens, scholars, researchers and communities to be more informed about matters of public importance, and empowered to communicate more effectively and participate more actively in setting public policy." As examples she proffered her experience with the Midpeninsula Regional Open Space District, which was unable to allow the public onto the Bear Creek property to see it for themselves. "Keyhole is being used by a local group to perform virtual fly-throughs of the land, looking for potential access points and trails." But her real test case was in the process of unfolding.

At the time she was making the presentation, Moore was having a good close look at the seemingly occult proposal of the San Jose Water Company that in order to minimize fire hazard, they planned to log redwoods and Douglas firs on 1,002 acres right in Moore's backyard, with the assistance of the Big Creek Lumber Company. "We get this absolutely incomprehensible map in the mail," Moore told me, with a proposal to log "right at the drinking supply for one hundred thousand people in Silicon Valley, right by the reservoir. The terrain is so steep they proposed to log by helicopter. Let me show you."

Moore gave me a flyover on her laptop, showing the mountaintop community where she lives, which includes a day care center and an elementary school. "Can you imagine, helicopters, chainsaws, two hundred yards away from these kids? They submitted a four-hundred-plus-page plan. Nobody reads that! In which they don't say what they want to do is cut old-growth forest. Instead they specify that they will cut 'late successional forest' when they find 'defective trees' to sustain 'the health of the forest.' I mean, come on! Does this tree look defective to you?" She pointed out a gigantic redwood. "That tree is older than Christ—it's not defective. It has a split top because it's two thousand years old. They call this kind of tree 'defective.'" Despite the fact that she had long won this battle, Moore was full of emotion recounting it. She told me the redwood habitat is home to osprey, and to red-legged frogs that are currently endangered.

Next Moore showed me the proposed area to be logged overlaid in red. I gasped, and she laughed. "Everybody in the room made that sound when I showed this," she said, referring to a community meeting. The idea that taking the central chunk of forest out of rolling mountaintops of green was necessary for fire prevention looked like a mighty stretch. "Later we learned that the San Jose Water Company had tried to sell themselves to a German company but had failed to because it wasn't valued highly enough. Once you get a logging operation licensed into perpetuity here, you make it very valuable indeed." When confronted with this, the San Jose Water Company replied that they were obliged to seek the maximum value of their holdings for their shareholders. "So now we're getting some honesty here," said Moore, joined by a village full of support to stop the logging plan. She helped form a group called Neighbors Against Irresponsible Logging. Their mascot is Charley the Beaver, "because he's a responsible logger."

A key advisor with long experience in forest activism told her, "It's a terrible idea, but it's legal." The San Jose Water Company had submitted its logging ambitions under a nonindustrial timber plan intended for small landowners. To qualify for it, you can't own more than 2,500 acres of timberland. The company claimed they owned 2,002 acres.

"We struggled for two years," Moore said. She enlisted the help of Ken and Gabrielle Adelman, who self-finance and run the California Coastal Records Project. The Adelmans fly a helicopter and take aerial photos of the coast, which they make freely accessible to nonprofits for all kinds of environmentally and socially responsible purposes. Moore went up with the

Adelmans to get aerial photographs of the intended logging site. In spite of the wow factor of Moore's Google Earth visualization, it was the Adelmans' pictures, purposed in a fairly mundane fashion, that would win the case. Showing me the visualization, Moore explained that each element of it originates in a thirteen-megapixel photograph of the actual trees. A conservation biologist, Adelia Barber, identified every single tree in the photographs. The total tree count exceeded what would be possible within the acreage claimed by the company and disqualified it from claiming nonindustrial timber logging allowances. "We did this polygon by polygon, by hand. It took time. Now you can automate this kind of work with Google Earth Engine. (Google Earth Engine is a cloud computing platform that makes it possible to access and analyze satellite imagery at a scale previously not possible on a personal computer.) This is a real inspiration to me. With Google Earth Engine, the map of a forest is substantiated by photographs." If Google Earth Engine had been around when California was being surveyed and dispensed, nobody would have been able to claim a piece of land was rocky if it was actually not. It takes interpretive license out of the mapping process, and that's a good thing.

After Moore gave her presentation at Google, she was promptly offered a job. As the fight to stop the logging continued, Moore got an assist from none other than Al Gore. "He came to Google in 2005, and he asked Eric Schmidt [then Google's CEO], 'Are there any environmental applications of Google Earth?' And people said, 'Go talk to that new engineer, Rebecca Moore.' I was like, gulp." (But in fact, I believe Moore capable of self-possession in any company.) "I showed him a project I was doing with the Sierra Club and with the Natural Resources Defense Council. And then I said, 'I have this local example. We think this logging is going to be dangerous and the people involved aren't being open with the community. They say it's all about water quality and reducing fire danger but those explanations don't stack up. And everybody told us, the San Jose Water Company never loses. They will out-legal and out-everything us." Gore said he was familiar with the area and offered a letter of support. It helped Moore's group get positive press. But as the definitive hearing approached, the Google Earth visualization of the proposed logging was barred from presentation.

Adelia Barber, the conservation biologist helping Moore and others in Neighbors Against Irresponsible Logging, applied her own science to the aerial photographs taken with the California Coastal Records Project. As Paul Rogers of the *San Jose Mercury News* described it, "While other logging battles feature environmentalists sitting in trees, this one featured Silicon Valley computer programmers sitting at their monitors." Barber analyzed the more than seven hundred photos taken from the Adelmans' helicopter, circling areas of redwoods and Douglas firs she then measured exactly, with the help of software. "Barber showed that the San Jose Water Company actually owns 2,800 acres of forest," Moore crowed. "That was our silver bullet." To qualify for the nonindustrial timber plan they were proposing, the holdings could not exceed 2,500 acres. "Now it was considered a commercial operation," Moore explained, "and that meant all kinds of additional qualifications would have to be met to log it." The California Department of Forestry and Fire Protection denied the logging plan. The decision was appealed once, and struck down again. The land in question is now under serious consideration by Midpeninsula Open Space, which seeks to protect it.

Story Map
Soon after Moore joined Google she began to be approached by nonprofit activists who had noticed her logging victory and

wanted to know more about how she had done it. Her official job was technical lead for Google Earth Layers. When Google acquired Keyhole, it needed to stitch satellite imagery into a 3-D rendering made from topographical information—"they needed roads and borders and the names of cities on the maps," Moore explained. Poetically, Moore's technical day job was in integrating a vision of the terrain with signposts for its navigation. Google famously encourages its employees to spend 20 percent of their workweek pursuing a dream, and work that has come out of it includes Google News and Gmail; Moore thus began Google Earth Outreach as a passion project.

Today it is her full-time job, to which she has added Google Earth Engine. Essentially Moore and her team have aggregated forty years of satellite imagery of the earth, cleaned up a time-lapse synopsis of it so that the imagery reads cloud-free, and hosted it with Google's computing power. Part of the impetus to do so came from Moore's question: "We've handled space on the maps. But what about time? The fourth dimension—so important in global change." The satellite imagery of change over time existed but was basically inaccessible, stored safely away in a vault in South Dakota. The utility of unlocking this data is demonstrated in a project Moore and her team undertook with University of Maryland scientists to show global deforestation. Processing data from nearly seven hundred thousand satellite images on ten thousand computers running in tandem, the project produced images of global forest-cover change in a couple of days. "Part of this is that we all work at a place that takes the idea of 'big' very seriously," Moore demurred.

You can see the global deforestation as well as other very cool visualizations of change over time on Google Earth Engine's Web site (earthengine.google.com/timelapse). Check out the growth of Las Vegas as nearby Lake Mead shrinks.

Still sounding surprised that he was able to pull it off, architect Charlie Kleemann told me how, after he heard tell of Moore's Santa Cruz victory, he stayed up for forty-eight hours teaching himself how to use Google Earth and figuring out how it could help his cause. "I just knew there was something wrong with it," he said, referring to a proposed gravel quarry in his San Luis Obispo neighborhood, though the operation was very near to getting a green light from the local planning commission. Starting with a final site configuration of the quarry's application, Kleemann used a 3-D modeling computer program called SketchUp to visualize the quarry, and pulled it into a Google Earth map of the proposed location. "This is when the dialog started to change," Kleemann told me. Actually showing the dimensions specified in the proposal revealed inconsistencies and misrepresentations in the environmental impact report. "We live in a conservative community that preferences property rights," Kleemann said, "but this information was just so straightforward. The quarry would have ruined our town, impacting many, many residents."

This is a critical part of Kleemann's story. As with Moore's work in her own backyard, the success came not simply from albeit brilliant purposing of technology and mapping, but also from the fact that he had a coherent community to work with, neighbors who understood they needed to stand together to protect resources they all valued. At the same time, the technological offerings helped the community see what they shared together.

"The message is simple," Kleemann told me, referencing the initial pass his community was prepared to give industry. "Yeah, it's a stacked deck, but there are tools out there for you to use to take action. You've just got to step up and do it."

CHAPTER SIX

Into the Woods

Emily Burns is a very fine name and suits the person who bears it, but I persist in wanting to call her "Fern." This might in part be because, since 2012, Burns has been running Fern Watch, a citizen science project she rather brilliantly conceived to track climate change in the redwood forests along their native range, from the Oregon border down to Big Sur. Burns is fresh-faced and petite, and she told me that one day back in 2008 she was making her rounds in the Humboldt Redwoods and registered the obvious: proliferant western sword ferns towered over her head. "Down below Santa Cruz they come up to my knees," she said. Redwood forests are not just made of trees, but include many associated species, especially western sword ferns. "Ferns are bigger and there are more of them where there's more water," she told me, "more rain and more fog. In a dry year, the ferns will show stress that is much harder to detect in the trees themselves."

"Aren't redwoods a relictual species?" I asked her one day, which was something of a taunt. Redwood trees are of an ancient lineage. In the days of dinosaurs, redwoods ringed the whole planet. Today three species of redwood survive, the giant sequoia, *Sequoiadendron giganteum*, which is the heaviest of living things; the coast redwood, *Sequoia sempervirens*, of which members are the tallest of living things; and the dawn redwood, *Metasequoia glyptostroboides*, a relative of the previous two found in the United States and which was thought extinct but rediscovered in China in 1944. Today the historic range of the giant sequoia and the coast redwood is limited to a fog belt along the West Coast. On the one hand, things look very grim for the native range. The milieu these trees have been adapted to for hundreds of millions of years is on speed dial for some kind of face-lift, probably a botched one, thanks to climate change. A certain sort of scientist makes the observation that it's kind of miraculous these giants have lasted this long here and maybe it's time for them to go. This perspective brings outrage to the pretty face of Emily Burns. "These trees have a fierce life force," she retorted. "They have survived conditions we can only guess at. I don't think we should just write them off."

Burns is mostly a cool scientist, but when her morality and commitment come to the fore I'm consistently reminded of the opening line to *Charlotte's Web*, by E. B. White: "Where's papa going with that ax?" I'm convinced that Fern Arable, the character who memorably preserves the life of a certain pig named Wilbur, grew up to become Emily Burns, PhD, science director of Save the Redwoods League.

A couple of years ago I followed Burns with a small group of citizen scientists into the East Bay Regional Park around Chabot Space and Science Center in Oakland, California, to take the measure of ferns therein. It was a rainy day, but once under the trees' canopies we were splashed only by an occasional droplet, and these seemed to be coming from the trees themselves rather than any clouds above them. The forest around Chabot is pretty easy to walk through, the ground carpeted with fragrant fallen leaves and duff. It is a tidy woods. Even as I was having the thought that the space between the trees seemed cleared for an unfolding fairy tale, we passed a ship-like wooden structure probably built for that purpose. "People are always trying for something primordial in here," Burns remarked. Just a few days before, she had accompanied a reporter and photographer in here to document the last remnants of a historic stump. Reportedly, this was an especially tall redwood that had been used by ship captains coming into the bay as a sight line to help them avoid crashing on the so-called Blossom Rock in San Francisco Bay—before the tree was felled in the gold rush fever for logs. Burns told me they came upon a bevy of naked women in nymph-like poses arrayed along a fallen redwood trunk. "The male photographer was fully clothed," she drily remarked. "It was so embarrassing, I kept my head down as we walked past them!"

Ironically, people can stage their fantasies in here partly because this is at least a third-growth forest. A month or so later I went with Burns to the Grove of Old Trees in Sonoma County, a twenty-eight-acre remnant forest, which has never been logged and has been protected since 2000 by Save the Redwoods League. The trees are staggering, but the ground is a dense tangle of vegetation. Nobody's posing naked in there. Whereas the ferns in the East Bay redwoods seem judiciously placed, in the Grove of Old Trees they are in many cases obscured by other plants, including small trees, and measuring them is a bit of a trick. It's hard to tell where one fern ends and another begins.

The East Bay redwoods were mercilessly logged several times. By 1893, Dr. William Gibbons, one of the medical doctors who was an early member of the California Academy of Sciences, was already doing historical ecology to discern the outlines of what had once been hundreds of acres of redwood trees. Traipsing among a "sea of stumps," he discerned "very substantial hints of a very ancient history of this patch of forest overlooking Oakland and now so nearly extinct." He observed that the "hidden depths over which we build securely our villages and cities hold records yet to be unfolded."[1] Gibbons showed the remnant forest to John Muir and Alfred Russel Wallace, who both decried its demise. That didn't stop it from being logged again once the subsequent generation grew up tall. As would seem to be the perpetual case, when developers want to build something, the forest becomes timber.

Burns made an early academic splash before she quite had her PhD in hand. One of her professors at UC Berkeley, Todd Dawson, had shown that in addition to soaking up the moisture from fog through their roots, redwood trees make substantial use of it directly through their leaves. Burns set about studying whether the same might hold true for the plants associated with redwoods. Her 2009 paper "Foliar Water Uptake: A Common Water Acquisition Strategy for Plants of the Redwood Forest" shows that, yes, the same fascinating modus operandi exists among eight dominant plants in the system, including ferns, shrubs, and broadleaf and coniferous trees. "I didn't think that it was going to be unique to redwoods," Burns said. "But I had no idea it would be so common." All these species absorb rain, dew, and fog water directly into their leaves or, in the case of the trees, through their crowns.

The phenomenon goes a long way toward explaining how redwoods have persisted in California's Mediterranean climate, with our protracted dry spells. The answer is: fog. One reason we still have redwoods is, once again, that oceanic upwelling, the same off-coast event that brings a fantastic panoply of sea life to feast around the Farallon Islands. Wind currents blowing offshore between March and September kick deep, cold water (full of nutrients for the kelp, inverts, fish, and marine mammals), and when the cold hits the hot summer air, the result is fog. Fog is held to the coast by mountain ranges that prevent its immediate dissipation, and redwoods and their vegetative ilk grab ahold of it. Redwoods intercept enough fog with their massive crowns that it provides 20 to 40 percent of overall precipitation required by the tribe.

Fog is generally cold, but it's also a proverbially hot topic in research on climate change impacts because of this capacity to compensate for lower rainfall in places like California. But the complexities of fog are just beginning to be unpacked. Research published by Dawson and James Johnstone in 2010 shows that fog has decreased by 33 percent over the past century, and trends continuing in this direction could stress the redwoods.[2] At around the same time, another set of researchers evidenced fog increasing rather than decreasing, due to the hotter temperatures brought to us courtesy of climate change.[3] The apparent contradiction is nobody's mistake but reflects the complexity of when and where you study fog. One place to find a more definitive guide to fog impacts on redwoods is in the trunks of the trees. Dawson's research today is focused on using isotope analysis to discern the markings of fog versus rain, hot years versus cool, right in the substance of the carbon and oxygen atoms that make up the tree's cellulose.

In 2009, Chabot Space and Science Center approached Burns to help develop a citizen science project. Chabot is one of twelve science institutions around the country collaborating on a project to better communicate climate change impacts, and according to Burns, "Redwoods were the iconic obvious habitat." She suggested tracking sword fern response to weather over time, partly because it is so easy for people to grasp that, yes, a bigger sword fern is getting more water than a smaller one. Armed with the field biologist's usual arsenal of tape measure and clipboard, Burns monitors transects in ten redwood forests with the assistance of her education and interpretation manager, Deborah Zierten, who brings hundreds of schoolkids traipsing into the forest each year to help.

Burns, Zierten, and a shuffling cast of citizen scientists and interns count up fern fronds, measuring them and noting evidence that insects have dined out here or not. Those little spots on the underside of some fronds are spore sacs and indicate fertility: I did not know that! So far Burns' data shows some change in the ferns in some places. In the Oakland site, nothing statistically significant. In other places, she's found a huge drop in size and health of ferns but said, "We can't pinpoint the cause yet. Dryness is part of it. Herbivory is part of the equation too, though. Caterpillars are breaking the ends of the fronds, so when the fern matures, it can't grow as much as it could otherwise. Maybe we're looking at natural ebbs and flows. Maybe the drought is affecting the herbivores more than it is directly affecting the ferns." Burns' questions about her data explain why she needs a lot of it over a long period of time, so that discernable patterns can emerge.

"The beginning of the citizen science project was a pivot point for me," Burns said. "I switched from just having an academic perspective to thinking about effective dialogue with the public. How could my research be more applied—how can I help land managers in the face of climate change? I think this focus made me competitive for the league job." Burns started in 2010 and dove into the League's Redwoods and Climate Change Initiative (which had begun in 2009). She brought the Fern Watch project along with her, and started another one for good measure, Redwood Watch. This ambitious program has been running for several years, with a substantial assist from Rebecca Moore's Google Earth Outreach. Whereas Moore was inspired by looking out from a mountain peak, Burns perhaps finds exposure most often by looking upward to the top of a tree that is literally out of sight. Still, with Redwood Watch her ambition matches Moore's global perspective. She wants to know where each and every redwood tree lives today all around the globe. And she wants each and every one of us to help her count them, which anyone can help do by contributing to the Redwood Watch project on iNaturalist.org.

Through the Trees

In 1917, a paleontology professor from the nascent University of California, Berkeley, had something of a flow experience, in which he apprehended the past as if he were watching it unfold into the present, while gazing at a redwood forest in the Eel River basin in Humboldt County. I have used this term to describe my own feeling of focus, energy, and surrender to a moment that is emphatically present yet informed by both past and future. John C. Merriam had his woo-woo trip looking at redwoods, having traveled with two other scientists to check out "a forest wall reported to have mystery and charm unique among living works of creation" and which was threatened with logging.[4] Merriam had just been piecing together the history of the Pacific coast, associating invertebrate and plant fossils with "a stratified geologic column." He was among the first to sketch out the natural history of the West Coast of North America within a world pattern of change over time—or evolution.

In the redwood trees Merriam saw a journey to his current moment in time of more than fifty million years, at which point redwoods were growing all over North America. Twenty million years before, redwood distribution had withdrawn to the Pacific coast from Monterey Bay to the Oregon border. Merriam registered history in motion. "In swift panorama the history of these trees . . . passed before me, stage after stage from the remote past."[5] With three other Berkeley professors Merriam quickly moved to protect these icons of evolution from logging. All three were members of the Boone and Crockett Club, a bastion of Teddy Roosevelt–style Republicanism that combines a love of hunting with a consequential commitment to conservation. They were significantly helped by another Boone and Crockett member, Stephen Mather, the first director of the National Park Service (founded in 1916). Mather had made a personal fortune building the company that sold the "20 Mule Team Borax" cleaner, which is still on the market today. With William Kent, who later bought and protected Muir Woods in Marin County, Mather put up the funds to make the first purchase of the Save the Redwoods League. Mather was bipolar and periodically hospitalized for depression. His passion for nature partly stemmed from the relief it gave him from personal darkness.

Who knows why—maybe because our brains are crowns at the top of our spinal cord trunks, maybe because we have a special affinity for fellow species that stand vertically—the tree is just about the most central, ubiquitous, transcendent point of reference in the natural world for *Homo sapiens*' mythologies attempting to describe the context and meaning for why we are born, live, and die. The biblical creation story has a central character and it's a tree. Darwin's sketch of evolution by way of natural selection limns a tree, and his progenies with PhDs are occupied right now filling in the blanks on the phylogenetic tree that in some perfect world will articulate all relationships among all life-forms past and present. Buddha had his enlightenment experience under a tree. Christ was crucified on a cross, the shape of which makes symbolic reference to the cycle of the sun.[6] It is significantly made of wood, which is grown by way of photosynthesizing rays from said sun. The cross is the physical juncture at which the alpha and omega of the Christian God intersects with the human bound to mortality. The cross is a tree. And the redwood tree is pretty much the ultimate tree model.

John Merriam was a scientist, but he was also fairly besotted with the natural world, with one author calling him the "philosopher of the redwoods."[7] Not only for Merriam but also for his vaunted colleague from Harvard, the botanist Asa Gray, the redwood was a testament to evolution, proving, in Gray's declaration before the American Association for the Advancement of Science in 1872, that the giant sequoia, *Sequoiadendron giganteum*, refuted Genesis—basically by evidencing life on earth millions of years earlier than the Christian origin story was first told. Darwin's *On the Origin of Species* was still relatively hot off the presses. But Merriam wasn't throwing spirituality out of the picture, not at all. He liked to repeat a quote attributed to Goethe: "Nature is the living, visible garment of God."[8] In the spirit of earlier endeavors to create botanical gardens that reflected or reassembled Eden, Merriam saw the study of nature to be at once the fount of classroom-style education and a revelation of deeper knowledge.

Redwoods continued their role in helping articulate earth history. Merriam's one-time student Ralph Chaney, eventually succeeding him at UC Berkeley, further reconstructed the redwoods' historical progression around the globe, positing that one hundred million years ago most of Western Europe, Russia, China, and Japan were covered by a redwood forest. This puts the redwoods around when dinosaurs roamed the earth. From Europe and Asia, redwoods proceeded to migrate west.

Chaney postulated a close relationship between the sequoia of eastern Oregon and the flora of China, and in 1930 the

American Museum of Natural History sponsored an expedition to St. Lawrence Island, which is smack between Alaska and Russia. Chaney sent the news to Merriam: "Sequoia flora of Middle Tertiary Age just received from St. Lawrence Island represents first actual record of land bridge between North America and Asia."[9] The missing tree link helped establish that *Homo sapiens* had come to North America over what had been the Bering Land Bridge. Fifty million years ago, trees had migrated to North America when the continent was covered in subtropical forests including fig, palm, avocado, and mahogany. Pleistocene cooling and further weather changes resulting from the newly emerging Cascade Mountains drove redwoods to their slim strip of West Coast. "For some twenty million years," Susan Schrepfer writes "the trees have grown only where bathed by the fogs that sweep off the Pacific."[10] Such monumental revelations into human and earth history yielded by the silent but present story of the redwoods inspired national and California park system boosters to call these safeguarded places the "University of the Interior," and "open-air schools."

In the tradition established by Merriam and company at the founding of Save the Redwoods League, the organization has continued to sponsor research into the oldest and tallest of trees. Whereas Merriam once wondered where the redwoods had been, we are now wondering where they are going. As the redwoods' history makes clear, climate and topography have a lot to do with where they can persist. Now enter climate change. One of the most challenging questions faced by the league today is where to focus their conservation efforts. Should they keep trying to buy parcels of the old-growth forests that fill us with awe? These are the 4 percent left after logging took out the rest of the original two million acres. These trees come with a densely layered ecosystem populated by fungi, lichens, bugs, birds, and other plants and tree species. Taking out the red-woods eliminates relationships that have evolved over millions of years. Or should the league spend its limited resources on the younger forests—even those marking two centuries? These don't have the same complexity as their older siblings but might have better adaptation strategies for dealing with a quick-changing future. The conditions that helped create the older forests might not be replicable.

"This question isn't going to be settled anytime soon," Burns told me. "Some people say we will have a bigger impact on the redwood forest if we move into the younger trees. I have reverence and respect for the old trees and I always will."

Where Are You Going?

In August 2013, Dr. Healy Hamilton presented an update to the biannual science symposium hosted by Save the Red-woods League on her work with Burns to model climate change impacts on redwood forests. Hamilton is chief scientist and vice president for conservation science at NatureServe, "a network connecting science with conservation," which began in the mid-1970s with the establishment of a natural heritage program. NatureServe has been documenting rare and endangered plants in all fifty states (and today has international reach as well), and is a leader in geospatial expression of what lives where under what conditions. Starting with 115 years of weather station data, Hamilton compared seasonal temperature and precipitation patterns in recent years to a twentieth-century baseline. As we all know, climate change signatures are already marking the course of time, and Hamilton's work showed significant changes across the coast redwood range—most noticeably, the number of cold spring days has decreased. The southernmost part of the redwood forest in Santa Cruz is experiencing temperature increases in all four seasons, but the Humboldt and Del Norte redwoods are showing climate stability.

To make future predictions, climate modelers take historical occurrences of anomalous weather years with exceptional levels of heat, cold, rainfall, drought, and all of the above, and run those numbers to predict where the extremes are likely to be showing up more regularly in the future. The conclusion here is again that Santa Cruz looks like it will get toasty and dry and the lush vegetation farther north will be impacted less and may persist more or less as we find it today. "While we don't yet know how coast redwood forests and their many associated species will respond to these changes," Hamilton averred, "our results suggest where the most severe changes are likely to occur."

Hamilton is the first to point out the trouble with her sort of modeling. "We're stuck with it," she said, "because there is no other way to grapple with the future, and we have to make conservation decisions now. But of course these models are based on very large-scale landscapes, and they use a very broad metric, temperature and precipitation. We need to have information from the ground. We need to know where redwood trees are, and how they are doing." Hamilton envisions citizen science as a living, close to real-time map of where exactly species are, marked by latitude and longitude and what level of temperature and precipitation (including fog in the case of redwoods) they are experiencing at the moment. Redwood trees will die off in some places, and they will persist in others. People have planted redwood trees all over the world, and some of these places may be good future locations to cultivate new forests. If we can keep track of how redwood distribution is changing, we can better support the trees where they are doing best.

The future of the redwood forest is also discerned by looking into its past. Another speaker at the league's symposium was

Allyson Carroll of Humboldt State University, who has been creating tree ring data for redwoods dating back thousands of years. This work, she said, establishes "a baseline, or ruler, upon which researchers can study tree growth, climate, fire history, and even date archaeological structures." Some of the trees Carroll analyzed encompass "the years 328 to 2012 AD." Yikes. Carroll pointed out that 1580 was a "small growth" year and a freeze occurred in 1739. "You can see a consistently small ring across trees in 1924," she said, and she has correlated those marks with climate data provided by PRISM, a data-monitoring service located at Oregon State University. "The trees are definitely picking up a signal."

Stephen C. Sillett, famous with his wife Marie Antoine for hanging out in and measuring up the supraterrestrial life in redwood tree canopies, gave a drill-down on how you measure carbon sequestration in these giants. "We mapped all the branches in these tall trees," he said, "because we want to know how many leaves they have, what is their photosynthetic capacity." Among the trees measured were several "superlative" trees, including one that is 1,450 years old and has 1.12 billion leaves, and another tree, at 2,520 years old, that pushes the age of a studied tree back by three hundred years. Sillett mapped 12,600 branches by height and diameter. He used his own measurements and the tree-core data to measure how fast the trees grow at each stage in their evolution. One tree "blew our minds! It's putting on two cubic meters a year. Good God! It's fairly stable till age nine hundred, and then it just takes off." In general, the redwood forest has been growing better as fog is declining, which makes sense. They have more direct, unclouded access to sunlight absent the fog. But the trend is disconcerting. How long can redwoods go on with less fog, as rainfall decreases in much of its range as well? And what about those fog-adapted species on the forest floor? To make a visualization of the redwood forest to help Save the Redwoods League launch its Redwood Watch citizen science project, Emily Burns sent skeletonized models of branches Steve Sillett had recently published to designers at Google Earth. "We wanted to show users the difference between an old-growth forest and a younger one," Burns said. "We modeled Jedediah Smith"—a state park that has two-thousand-year-old trees in it—"and Mill Creek, a young forest with small trees close together. Mill Creek broke the servers—although the density of the biomass in Jed Smith is greater, so many more young trees in Mill Creek challenged Google."

Now, several years later, Google Earth Engine would be able to handle that data no sweat. The visualization is still a fun ride to take, with narration by Peter Coyote. Google "Finding the Redwoods of Tomorrow" for the video.

Redwood Watch and Fern Watch are among the most accessible of citizen science projects to join as an individual or family—go to iNaturalist.org and search "projects" for "Save the Redwoods League." Wherever you are in the world, and wherever you travel, be on the lookout for redwoods and document them! Tuning in to these mesmerizing trees with their ancient lineage evokes the long time scale over which they have migrated around the world. They link us not only to the past but to the future as well. Where they have been planted today and are thriving is an important signal about landscape resilience in the context of climate change.

CHAPTER SEVEN

We All Want to Change the World

The walls have ears, though they don't hear a thing. In late fall 2014, several hundred donors, trustees, scientists, and hangers-on such as myself listened as Dr. Jonathan Foley, newly named executive director of the California Academy of Sciences, stepped up to the microphone in the institution's Tusher African Hall. Tableaus of taxidermied specimens bore witness to words likely never before addressed to the annual fellows' dinner. "We are at an inflection point in history," Foley pointed out. "What we do or don't do decides the future of life on earth. You didn't choose this moment, and neither did I."

The old-fashioned African Hall dioramas exemplify the public floors of classic nineteenth-century natural history museums. The stuffed mammals are the only reference the building makes to its previous incarnation, razed to make way for the current Renzo Piano–designed monument, which opened in 2009. A window looks in on the academy's live African penguins, just standing there on their stairs or waddling around and making an amusing counterpoint to all the tuxedos in the room.

Alone among its peer institutions, the academy comprises three major components, the Kimball Natural History Museum, the Steinhart Aquarium, and the Morrison Planetarium. It has a multistoried "rain forest" and living roof. If grappling with this multivalent invitation to various perspectives on the natural world yet leaves the visitor short, feeling perhaps entombed in the building's caverns of concrete, there is respite—the Academy Café run by the Vietnamese restaurateur Charles Phan. Try the artichoke and wheat berry salad.

Foley is clean-cut and easy to listen to—articulate but not intimidating. He's a manly John Glenn with an extra skosh of wonk. Foley's main discipline, global ecology, is highly analytical and inclined toward the rocket science definition. His many accolades include the prestigious Heinz Award for work with computer models and global data sets that reconceptualize the intersection between land use, particularly agriculture, and the world's ecosystems.

"We did something nobody had ever done before," Foley told me at a later date, explaining the work that made him famous. One of the biggest drivers of climate change and general ecological degradation is how the world feeds itself. But how to get a handle on exactly where, when, and how the damage is being done, by the conversion of habitat, the insertion of massive amounts of nitrogen fertilizer into the ecosystem, and commensurately large expenditures of water? "Satellites can see green stuff—you can see different types of green—but it's hard to tell corn from soy unless you get clever at training computers to recognize different kinds of plants," Foley told me. "We realized that every country in the world does censuses of ag for tax purposes—county by county, with clipboards, people go around in trucks. Nobody had put all that data together for the whole world at the level of counties and districts. We're tracking thirty thousand political units—in some parts of the world, we track by country. We digitized the whole thing and used it as a first guess or calibration point for satellite data."

Foley set about reconciling how we read the satellite imagery with the ground-truthed information. "We'd get it completely wrong, we'd think we had corn somewhere and it turned out to be wheat, or in a different location than we were discerning from the satellites. So we went back and back and back and refined our models till we got it right."

Foley's methodology helps keep track of what is essentially ecosystem devastation undertaken in order to support human lives. Here is the conundrum at the heart of how we live today. There is no greater or more threatening contradiction. What you get is what you see—a self-consuming system. We're destroying the ground beneath our feet as well as the atmosphere swirling around our heads.

At the fellows' dinner Foley had mercy and showed us some slides of gorgeous nature but he also stepped up to the inflection point and presented some blasted landscapes. "We've already destroyed whole ecosystems," he said. "It's insane."

Later I asked him about how we are going to support population growth headed toward nine billion.

"Is that sustainable?" he said. "No. We are not living sustainably right now, at this level. If not a single new iPhone were to be made, not a single new housing development, we are still vastly over-taking from what the earth can provide."

Foley told me he thinks that we will get a handle on the situation—eventually. "It's going to take a couple of generations," he said.

"What about the plants and animals that go extinct in the meantime?" I asked. "What about the human suffering?"

"Yes," he concurred, nodding his head.

I felt big, palpable relief that one so willing to take on some of the world's biggest problems had come to helm the academy. At the same time, I was not alone in wondering what Foley's sort of science had to do with the taxonomy and systematics the academy's research had been based on for more than 150 years. I hoped he wouldn't throw the baby out with the bathwater. It was easy enough to look upon the traditional taxonomists as anachronistic, "stamp collectors" hewing to a naive and indulgent view of how to order reality. Taxonomy had been assailed before. I knew many of the academy's senior scientists were full of foreboding and muffled outrage that their discipline had been so thoroughly ignored in the choice (made by the board of trustees) of Foley to run the place. In the words of one senior curator at the academy, his work had little "to do with actual living things" but trucked in visualizations and statistical analysis.

Heading toward drinks in the academy's foyer after the fellows' dinner, I ran into Dr. Matthew James, a professor at Sonoma State University, an academy fellow himself, and an authority on the 1905–06 academy expedition to the Galápagos Islands. "I've come to watch a nineteenth-century institution turn into a twenty-first-century institution," he said, but didn't look very happy. Tonight a handful of scientists would become fellows of the academy—an honorary distinction of association that comes with minimal duties. And six of the longest-tenured scientists of the academy would retire. These men had made up the core of the academy for nearly fifty years. There had been upheavals and regime shifts before, but like limpets in the tide pool, these guys had hung on. There was a sense tonight that their run was over. James was right. A torch was being passed.

Then I ran into Dr. Michael Ghiselin. His bristly white mustache and goatee appeared to materialize ahead of the rest of him. He was almost unrecognizable in a tux, without his customary crumpled fishing hat or Members Only jacket. Ghiselin was my longest-time friend at the academy—we had worked together in 2009 on a San Francisco citywide celebration of Darwin's birthday and publication of *On the Origin of Species*. In addition to being a Darwin authority, Ghiselin had long studied gastropod opisthobranchs, a.k.a. nudibranchs, and in these spineless wonders he discerned a major new idea about hermaphroditism. Why do some fish and other animals change sex, often starting as female and then becoming male? Riffing on Darwin's ideas about sexual selection, Ghiselin proposed that males fight for the opportunity to reproduce, and females that get big enough to win the fights switch sex. OMG! This idea spawned (shameless pun) a vast literature on reproductive biology. One of his main ideas is articulated in his book *The Economy of Nature and the Evolution of Sex*, which he described as a cross between the Kama Sutra and *The Wealth of Nations*. As to how things work in nature, Ghiselin sees things as do the pundits in presidential elections: "It's the economy, stupid."

If we place Foley on one end of the spectrum of science today, Ghiselin is on the other. Science, in Ghiselin's view, is not about solving real-world problems. It is about "pure" inquiry and is undertaken without a goal. A field like Jon Foley's global ecology is welcome to be "applied"—to approach real-world problems looking for practicable solutions—but not welcome to determine the research agenda of the academy. The scientists who daily peer through microscopes at actual organisms, physical trophies of the natural world as opposed to abstract data points on a grid, are looking at life with open minds. This is the only way to make discoveries, to make progress. Only in this spirit can intellectual honesty flourish. (Ghiselin's longtime colleague and friend Terry Gosliner, curator of invertebrate zoology and geology, has said to me, "Mike is a snob. But he's actually a citizen scientist. He's able to have an idea and pursue it and the academy supports him in that." I can just see Ghiselin's louvering eyebrows upon the suggestion he is a "citizen" of anything but the absolute truth.)

What inspired me in Foley's talk had irritated Ghiselin anew. In particular he hated the word *sustainability*. "What does it mean?" he wailed. Then there was that practically useless term, *biodiversity*. Ghiselin was right that these concepts are vaguely bandied about without real definition. Still. We have had this conversation before.

"Mike," I reminded him. "If there's no nature left, then there will be nothing to study."

He glowered at me. Ghiselin would probably agree that someone should definitely save nature, but that this is not the job of hard-core science. Nor should pure scientists be turned out of natural history museums so that global ecologists can take their places.

Another of Ghiselin's distinctions is as a philosopher of science. He's in fact preoccupied, if not obsessed, with answering the question, "What is a species?" This question was a major subject of Darwin's. Darwin and his cohorts parsed out evolution by natural selection by asking the questions of what creates difference between species, and why these differences exist. How do new beings come into being? The question goes to the crux of creation. A major way into it is by way of biogeography—or who lives where, how long they have lived there in what company, and the features of the place vis-à-vis climate, topography, and other physical attributes. The pure academy taxonomist today is looking at, say, a single family of plants and drilling down to the molecular and genetic levels on how various members are related to each other, and they are identifying when in earth history one branch of that plant's evolutionary tree sprouted into two and on and on. We know in broad strokes that a change in environment is one precursor that can provoke a species to split into new forms. Scientists determining exactly when and exactly where this happens are discerning the history of all life on earth and getting at its most fundamental mechanism, by which something new is created.

The devil is in the details, of course, but in one way of looking at it, the divide between the close scrutiny of life-forms and the purposing of massive data sets to figure out how various earth processes are interacting is a matter of scale. The big data sets need pieces of information with which to work. Specimens in the world's natural history museums are arguably the most valuable of all data points, the substance of life itself. They embody instances of life expressed in bodily form and they mark where this bodily form occurred and when. They provide us with the raw materials for assembling a picture of life now and then, a road map of how life made its way from the past to the present. This information is what we've got to go on in figuring out what is likely to happen in the future. Since climate change and other human impacts are causing mammoth and fast alterations of earth systems, we need this information to help us adapt to what's coming.

For today's pure as well as applied purposes, historical museum collections have some major weaknesses. Until fairly recently collection was undertaken without complete awareness that date and time are critically important to figuring out how life has unfolded, so most historical specimens do not include very specific metadata about when and where they were taken. Collections that comprehensively represent a region or an ecosystem are few in number. Another issue is digitization. Since scientists are asking themselves global questions, they need access to information from all over the world. There is an ongoing push to take photographic records of physical speci-mens and get them included in universal databases. But most of the specimens in today's museum drawers have not yet smiled for the camera. That means scientists have to request that physi-cal specimens be sent to them from the institutions that store them, which takes time and labor.

The academy's traditional taxonomists have wrestled in self-scrutiny on these issues for a long time. How to do faster, more efficient exploration of the world's yet-unknown species, and how to get that information more quickly and easily avail-able to other researchers? Given the strengths in the academy's collections and the expertise of many of its researchers in differ-ent parts of the world, where should major efforts be focused? In recent years the academy has sent annual expeditions to the Philippines, continuing to explore an area represented in its col-lections over the course of a century. In 2014 the focus was on Verde Island Passage in the Coral Triangle, a place so dense with diverse life-forms, many of them never before seen, that those normally reticent about such things are calling it "the center of the center" of biodiversity.

The academy's work will inform conservation decisions in the Philippines, but as important as that effort is, the institu-tion knows it is not enough. The academy has an authority and a presence in regional California as well as the continental United States that comes with a responsibility to help save nature here as well. As beguiling as exhibits about burgeoning life in the Philippines may be, such displays don't connect the hundreds of thousands of people who walk through the academy doors with nature closer to home.

In the summer of 2011, a core group of academy scientists convened six sequential meetings in which they dug deep into the purpose of their work and how they could make it connect more fully with the public. Writing that had been sketched in pencil for years was now writ large in Day-Glo on the Renzo Piano walls. The ivory tower days were a thing of the past: let's take our esoteric science and make it real. Jon Foley's predecessor, director Gregory Farrington, instructed them to make sure to meet three criteria: their science should be competitive with that of other prominent research institutions (and so attractive to funders); it should address relevant and important questions; and their research projects should have a strong connection to public programs. Eventually, the meetings led to the decision to establish a citizen science program at the academy. The concept grew organically in conversations that stemmed from the heart of the institution's purpose and direction.

Terry Gosliner led the charge. In making among the first or perhaps even the very first mention of citizen science as a potential frame for bringing academy research into the world at large, Gosliner mentioned at this meeting that an ongoing nudibranch survey he was involved with had enlisted citizen scientists to help get the job done. In the next meeting, addressing California biodiversity, the subject came up again. "We have the largest collection of California flora in the world," one participant remarked. "We need to resurvey all those historical transects to see what's still where and what isn't." Gosliner again noted that such a massive project would have to "rely heavily on citizen scientists. Native plant societies and other groups could be tapped for volunteers." The group discussed citizen science initiatives in New Zealand, which is a leader in embracing this brave new world of mob-sourcing. Mention was made of citizen science in the city of San Diego. "They've trained volunteers to document every square inch of San Diego County," one participant added. "There are models for this in other areas. It works well."

At the California biodiversity meeting, Gosliner seemed to muse aloud. "What's missing is a critical mass of leadership," he opined. "It's not enough to just put up stuff for other people to use," he said. "We need to identify the endgame." He remarked that it would be necessary to integrate a resurvey project with the research goals of academy scientists. "We need to have a critical mass of people thinking of this as their *primary* research . . . not a side project."

Circling back to California itself, as one of the richest biodiversity hotspots in the world, the conversation really hit home, so to speak. As the academy possesses the most comprehensive collection of California biodiversity, it has the opportunity to be the best in exploring, understanding, and developing strategies for preserving our state's unique biological heritage.

Eventually two test cases were settled on. One, now completed, was a survey of Mount Tamalpais in Marin County, in partnership with the Marin Municipal Water District, which owns a great deal of the mountain. The other is the ongoing monitoring project of the intertidal area in Half Moon Bay in conjunction with the Greater Farallones National Marine Sanctuary and the Fitzgerald Marine Reserve. In May 2012, the academy convened a big citizen science powwow, and over several days many of the country's leading practitioners of such testified about what they do and why, what works well and what doesn't, and pondered future directions for saving the world. The most significant step the academy has yet taken has been to incorporate iNaturalist.org into the institution. This is a social-networking-style Web site and app whereby anyone anywhere can make contributions to species lists. iNaturalist is, yes, a great outreach tool for public programs. But it is also the superhighway to amassing digital records that can be purposed by big-data science. iNaturalist bridges the worlds of the old taxonomists and Jon Foley.

Clubby

We've been here before, in a different way. In opening its arms to the scientific participation of regular people, the institution is returning to its roots. The academy was founded entirely by amateurs. And the contributions of amateurs working at the academy have been fundamental to the working out of one of arguably the most pivotal scientific breakthroughs of all time, the theory of evolution by natural selection.

On April 4, 1853, a group of "seven San Francisco gentlemen," in the words of historian Theodore Henry Hittell, met and declared their interest in organizing an institution "for the development of the natural sciences."[1] A report enumerating the reasons for this asserts, "We have on this coast a virgin soil with new characteristics and attributes which have not been subjected to a critical scientific examination." A quick glance from "the eye of the naturalist" sees "a field of richer promise in the department of Natural History in all its variety than has previously been discovered." The report hits on three main objectives for establishing a museum of natural history: to elucidate nature in general, to further democracy, and to make money. Revealing the hidden mysteries of nature would contribute to the progress of the age, by their application to farming and commerce. The language of a new and proud imperial democracy permeates the proceeding notes. Science was making our country "the envy and terror of despots everywhere," and America's "avenues" of wealth would help spread the blessings of our "free institutions" to the ends of the earth.[2]

It's worth remembering that abolition was in full swing in the 1850s and much of the United States still clung to slavery, raising the question of whom "free institutions" were really for—clearly not everybody. On top of that, while the California Indians were originally abused and their cultures torn apart by the

Spanish mission system, the gold rush turbo-charged the stealing of their land and the outright murder of the people. Not a freedom song. As for money, Hittell remarks, "Notwithstanding the flush times…there was substantially none for the Academy." He records the secretary's notes for the first serial meetings indicating, "No money received at this meeting," "No cash received," and the "shorter, though quite as expressive, phrase 'No cash,' which had already been used several times." The academy, however, "was born to live, and it manfully struggled on."

That the magnificent seven who founded the academy were "gentlemen scholars," in the words of botanist Barbara Ertter, and not exactly scientists is perhaps evidenced in the way they got going.[3] The first issues of the academy's proceedings noted the donation of a living owl, "caught near Point Jackson on San Francisco Bay."[4] The next meeting's update recorded "that the owl was lost." An "extensive assortment of plant seeds" was donated by Dr. Albert Kellogg. The happy sum of twenty-two dollars in membership fees was recorded on November 7. On November 28, "Dr. T. S. Anderson presented the Academy with a collection of plants, woods, and shells from Monterey and Santa Cruz and also from Rio de Janeiro, Valparaiso, and the Samoan Islands.…Dr. Wm. P. Gibbons spoke about the rocks of Telegraph Hill and the coast of the Bay."

Five of the seven founders bore the title "Dr.," but only one of them was a practicing physician. The above-mentioned Dr. Kellogg had a pharmacy, but he "was almost too much engrossed with hunting and working over new plants to patiently wait upon customers," according to Hittell, and he was "particularly fond of trees."[5] Two of them were affiliated with geological surveys, one was an attorney, and one a real estate broker. None of them was what we would consider today a professional scientist, but this was not unusual in the early days of natural history museums.

The academy founders met weekly in a rented room and initially published their scholarly findings in local newspapers. In 1855, the library of the academy totaled sixty-five books, and all the curators shared a single cabinet. The road to professionalism was very gradual, and overall, the gentleman- and gentlewoman-scholar model persisted well into the early decades of the twentieth century. Women were expressly welcomed in a motion initiated by Dr. Kellogg in the academy's first year: "Resolved, as the sense of this society that we highly approve of the aid of females in every department of natural science, and that we earnestly invite their cooperation."

As the institution evolved, its identity reflected, influenced, and was intertwined with the births of the University of California, Stanford University, state and federal agencies including the Geological Survey and the Forest Service, and advocacy organizations like the Sierra Club, the Sempervirens Fund, and Save the Redwoods League. Dr. Hans Hermann Behr brought the first whiff of true professionalism to the entity, joining the ranks in 1854. An aristocrat trained as an entomologist, a friend of Alexander von Humboldt, Behr made his mark as a botanist. His 1884 monograph *Synopsis of the Genera of Vascular Plants in the Vicinity of San Francisco, with an Attempt to Arrange Them According to Evolutionary Principles* in fact explains the concept of adaptation and speciation more clearly than many contemporary attempts. New species arise out of older forms, a process that "takes place by the individuals adapting themselves to external conditions, brought on by gradual geological changes, and it is astonishing what a variety of differences can grow out of the same old stock." Behr tells us that adaptations will continuously emerge "because in this world changes are constantly going on. There is neither pause nor return." The tone of his day and the bridge he was helping to build poke through the text as here he completes his thought: "The Creator never repeats himself."[6]

This last bit is worth taking special note of: the ability to distinguish the "new" from the previously discovered. The amateur botanists at the academy were aware that the California flora was special and unique and that they had a fantastic new panoply of species to introduce to science. But with so few publications with which to reference their findings, and with virtually no authority upon which to base their declarations of what was what, they were kind of stuck. East Coast and European scientists, considering the academy crew amateurs and upstarts, resisted species designations coming from California. But in this they were not always quite on the up and up. As Vladimir Nabokov wrote in his poem "On Discovering a Butterfly": "I found it and I named it... and I want no other fame."[7]

Naming new species was and continues to be the natural sciences equivalent of the Hollywood handprint, and those snobby professionals willy-nilly named the California species themselves. The academy battle for scientific authority went on for decades and finally received a critical blessing from Asa Gray. Gray famously founded the herbarium at Harvard University, and he was a friend and critical supporter of Darwin's. "It had been the custom of some Eastern men to describe all sorts of California plants from any kind of specimens, without ever having seen them grow," Hittell wrote. "They had not infrequently received credit which should have remained in California." Gray's encouragement probably helped bolster botany as the most vibrant discipline at the academy in its early days.

Natural Theology

Asa Gray is also an interesting figure in the shifting definitions of and relationships between the amateur and the professional.

Throughout the nineteenth century, botanizing was all the rage, an activity undertaken by hundreds of thousands of people of all ages, male and female, rural and urban. It's hard to imagine now, and if plant-focused citizen science projects today ever reach the historic numbers of hobby botanists, we will be in good shape (as long as they deliver their data to scientifically vetted databases). Botany was vaunted as an activity good for body and especially for soul; it was looked upon as a means to studying God's creations. Gray was probably Darwin's most important supporter in the United States, but he also held to the conviction that plant life reveals divinity. Quoting Matthew 6:28 at the opening of his textbook *Botany for Young People*—"Consider the lilies of the field, how they grow"—Gray put a fine point on it: "Christ himself directs us to consider with attention the plants around us." In general he supported the idea that natural selection and adaptation are part of God's design. Hewing to what's called natural theology, some botanists were very specific about how plant life reflected, as Elizabeth B. Keeney puts it, "God's continual orchestration of the world," citing, for example, its transformation of the environment to make it habitable. One botanical writer advocated gratitude for "the wisdom of the system."[8]

Gray eagerly enjoined armies of amateur botanists who supplied him with specimens and treated them as colleagues, though not without a personal interest. He was busy compiling a hugely ambitious *Flora of North America*, and among other things, he needed the California information to fill it out. It behooved him to maintain positive relations with the crew at the academy. Gray also cultivated the steady flow of California species identifications from the eventual California Geological Survey to Harvard, where the first set of botanical collections made by the survey were deposited at the Gray Herbarium;

they remain there to this day. (A duplicate set established the University of California's herbarium.) Gray thus paradoxically undermined the academy's efforts at "autonomous publication" and contributed to the premature curtailment of the Geological Survey itself by the California legislature, one argument against it being that it was serving Harvard more than the state itself.[9]

Despite Gray's interest in its contributions to the science (and thus religious implications) of botany, the survey was established for less lofty reasons. The gold rush was just that, hard and quick, over fast—like the saying has it, a flash in the pan. By 1860 the emergent state of California was looking for new ways to harness and monetize its natural resources and thus proposed a geological survey on the model of many older states, as "a hallmark of enlightened state administration. . . . The means whereby exploitable resources might be cheaply located and advertised to would-be investors."[10] The survey was dispatched to map the state and inventory its plants, animals, rocks, soils, and minerals. It chugged along in fits and starts, always begging funding from the California legislature, until the plug was pulled on it in 1868.

The academy had close intersections with both the survey and the nascent university system as well (which it still does), and in those early years there was something of a push-me-pull-you between the institutions. First was a tussle over who got the specimens from the decommissioned survey—the university did. The academy's ranks were at somewhat of an ebb at this time, and the regents of the university actually proposed incorporating the academy into its own, newly formed ranks. The director of the academy at the time, James Cooper, who had been a survey zoologist, fulminated against being "swallowed up in that Asylum for rebel Professors." (And the 1960s were almost a hundred years away!)

True Believers

The academy established itself at the biggest juncture ever in the history of the natural sciences, surely, and some would say in the history of all human intellectual life: at about the time that Darwin's theory of evolution by natural selection humbly began to disrupt centuries of belief about life on earth. Founded in 1853, the academy was a mere five years behind the presentation of Darwin and Wallace's papers on the subject of natural selection at the Linnaean Society. Included in the papers was an 1857 letter from Darwin to Asa Gray, elucidating his ideas. Darwin's biographer Janet Browne says Gray was "by far the most prominent botanist in the United States, and on an intellectual par with Louis Agassiz, his zoological counterpart at Harvard," and paints a picture of Darwin venturing his ideas before the august Gray with trepidation. Darwin worried that the devout Congregationalist Gray would "despise me" when he shared his ideas, but Gray encouraged Darwin: "Can you get at the *law* of variation?"[11]

Asa Gray teamed up with Joseph Hooker, the botanist to whom Darwin had first confided his ideas about "transmutation" ten years earlier, and the two scientists introduced the work of Darwin and Wallace to the Linnaean Society. Browne says they were subsequently "relieved to be able to talk about natural selection in public." The protracted tension Darwin and Wallace's ideas would create in science began right here. Gray outlined natural selection at a Harvard University science club meeting in 1859. He confessed to playing the provocateur, "maliciously to vex the soul of Agassiz with views so diametrically opposed to all his pet notions." Agassiz had become one of the earliest honorary members of the California Academy of Sciences, upon his acceptance of this position in 1854, Hittell writes, "requesting to be furnished with all the documents concerning the discovery of the viviparous fishes."[12] As Gray was keen on getting the new flora from California, Agassiz was keen on the fauna.

Browne says that in 1858 and 1859 Agassiz "dominated American intellectual life. He was well known as believing that all living beings, including humans, were created by divine fiat. Species were thoughts in the mind of God, he announced in his *Essay on Classification*." Agassiz reasoned that no classification system could be possible if species were always changing; he opined that since what would eventually be called ecology is a series of densely intertwined relationships, species must be constant. Darwin greatly admired Agassiz but this idea of the immutable thoughts of God was "utterly impracticable rubbish."[13]

Agassiz's antipathy and Gray's enthusiasm toward Darwin probably added some tension to faculty meetings at Harvard. But both men continued to be highly esteemed and their influence felt all the way to California. In 1872 Charles Darwin was elected an honorary member of the academy (a distinction conferred in absentia—he never visited here). The same year, Agassiz addressed the academy, speaking to an eager crowd big enough that the meeting had to be held in a venue more capacious than the academy's Clay Street digs. Agassiz praised the academy for hewing to more lofty concerns than the frantic gathering of gold and emphasized California's duty to increase scientific knowledge.[14] It was no small matter that America's premier intellectual stamped the mission of the academy with his imprimatur. Agassiz took the academy seriously and called it to a higher sense of destiny.

Rollo Beck's Birds

In 1903, a year after Leverett Mills Loomis became director of the academy, he noted that the museum's housekeeping had duly been set in order, and the institution was "now in readiness for exploration and larger accessions." Loomis wanted to put the academy on the map, as it were, of important science institutions, and the way to do this was to capitalize on the academy's relative proximity to the Galápagos. Thanks to Darwin, the Galápagos was the very locus of evolutionary thought. There was widespread eagerness to establish or refute the concept of evolution by means of natural selection—professionals and amateurs were on the case. Darwin's idea was that small variations in features could eventually lead to new life-forms. To measure and compare sameness and difference required large series of the same species, and scientific expeditions now trained their intentions on collecting these.

The Panama Canal was not yet open, and for the academy's main brethren, the Philadelphia Academy of Natural Sciences, the Smithsonian, or the American Museum of Natural History, to mount expeditions there would mean traversing half the globe. Loomis hatched a plan to mount the world's most ambitious expedition thus far to the Galápagos. Most of the collectors he assembled to staff it were amateurs. Only one, Alban Stewart, had an advanced degree, a master's in botany; his evidently prissy personality earned him the nickname "Granny." Rollo Beck, the expedition leader, was thirty-five years old, and J. J. Parker, the schooner's navigator, somewhere in middle age. The rest of the crew were all under thirty years old. Several were still teenagers. "Will Cruise South to Study Herpetology, Botany, Ornithology, Entomology, Conchology and Other Ologies" read a June 27, 1905, headline in a San Francisco paper.[15]

Al Leviton, one of the "retired" curators who yet puts in more than eight hours a day, working mostly as a historian of science and also as a herpetologist (he named a new species of reptile in 2012), has been at the academy long enough that his tenure overlapped with that of Joseph Slevin, who as a twenty-four-year-old was one of the eleven men sent by the academy to the Galápagos in 1905. Leviton remembered learning collecting techniques from Slevin. "Think about it," Leviton told me. "It was only seventy years after Darwin had been there himself! Darwin's *Origin* came out in 1859—what is that, forty-six years earlier?" Leviton remarked that at the time he thought of both Darwin's trip and the academy's subsequent expedition as historic events, "but not that removed from everyday reality." The academy has digitized the logs from each of the explorers, and Slevin's is by far the most narrative and captivating. Of hearing the news that life back home was never to be the same, Slevin recorded:

> We made the anchorage at Villamil at 2:30 P.M., letting go anchor in four fathoms of water. When the schooner swung to her moorings she was too close to the reef, so we got out the kedge and hauled her off a little. The Ecuadorian gunboat Cotapaxi was at anchor, and from her we got a Guayaquil paper announcing the fact that there had been a terrific earthquake in San Francisco and that the soldiers had been ordered out on the streets. This was the first and all the news of the great disaster that we had for some time. The Cotapaxi sailed at 3:30 P.M. for Guayaquil via Chatham Island. One of her crew died of yellow fever today and was buried ashore.[16]

Master and Commander

But the star of this particular journey is one Rollo Howard Beck. Master of the schooner *Academy* and leader of its expedition to the Galápagos Islands, citizen scientist exemplar Beck became one of the best ornithological collectors of all time. In 1918 he penned a five-page summary of his life till then, beginning: "On the 26th of August 1870, in the town of Los Gatos, Santa Clara

Co., California, was the announcement made of my entrance into the light of day. . . . My school days did not quite last long enough for me to graduate from the eighth grade in the Berryessa grammar school."[17] Beck worked in the fruit orchards around Berryessa and soon came under the tutelage of one Frank H. Holmes, a neighbor who had formal ornithological training and showed Beck how to identify birds, prepare skins, and mount specimens. When he was fourteen he made his first contribution to a museum collection, a common nighthawk (*Chordeiles minor hesperis*) sent to the Smithsonian.

In 1894 Beck joined the American Ornithologists' Union (founded in 1883), and the newly formed Cooper Ornithological Club, both of which have international relevance today, especially through their scientifically focused journals *The Auk* and *The Condor*. Through these affiliations Beck corresponded with senior ornithologists who were eager for his information on California species. He collected regularly, in the Sierras, in Yosemite Valley, in Santa Barbara, and in the Channel Islands, making significant contributions to science along the way, including the first eggs and nests of the hermit warbler and western evening grosbeak. In 1897 he collected for the academy for the first time, and his reputation grew.

That year, Britain's Lord Walter Rothschild funded the Webster-Harris Expedition to the Galápagos, instructing his employees "to outdo expeditions of Darwin, Baur, Agassiz and others" in collecting volumes of specimens from the islands, and he told them to do it with "NERVE, BACKBONE AND ENERGY."[18] Death, desertion, and general woes beset the original crew, which eventually regrouped in San Francisco. They were in need of an experienced collector and seaman to take the helm of the expedition, and Beck was the man for the job. The successful trip included discovery of the Galápagos flightless cormorant. While the bird's ancestors very likely arrived on the islands the birdlike way, on wing, over time the species shed the ability to fly. Lack of predators to elude meant it could get rid of what biologists call an expensive adaptation—the ability to fly requires an expenditure of energy, which means the bird has to eat more. Finding the flightless relative of a well-known bird epitomized the excitement engendered by Galápagos flora and fauna, then and now, so directly testifying to alternative strategies for living under different circumstances and developed across different time trajectories.

Beck showed his mettle early on. C. M. Harris's diary recounts a day long on walking and short on water in pursuit of the giant reptiles so beloved of Rothschild: "Two men each took a tortoise lashed to a pole and started for the coast. It was the hardest work I ever did for my part and I guess that the rest thought the same.... No dinner. No water." One of the other sailors was "completely exhausted"; by contrast, Beck "secured a rat."[19]

After the Webster-Harris Expedition, Beck went to the Galápagos three more times before 1905. He was mature, experienced with the islands, and a crackerjack collector. He knew where to find species and where to anchor the schooner. In putting together an expedition to the Galápagos, Loomis would have naturally looked to him to lead it. Beck was not only singularly capable of physically and mentally managing such an arduous undertaking, he was also adept at fine-tuned discernment of species. While Loomis wanted the academy's expedition to bring back an authoritative collection from every taxonomic group, his specialty was seabirds (the petrels, albatrosses, and shearwaters, in particular, which have uniquely tube-shaped nostrils), and at finding and collecting these Beck excelled. Beck devised a special toothed spoon for stripping fat from birds and a method for coaxing cryptic avians from beyond the horizon by "chumming." Like Hansel and Gretel, Beck would go out in a rowboat and drop bits of meat or fish into the water for miles and then row back and get the birds. He was able to attract even small birds that are harder to spot on the open ocean, like storm petrels. Field collectors had to be intimately conversant with flight patterns, color, size, and the geographical range of species in order to know what to collect.

All this talk of Beck's proficiency in "collecting" might get you to wondering—exactly how, exactly *what* is he doing to "collect" birds? And the answer is, mostly, he shot them. Jack Dumbacher, curator of ornithology at the academy today, told me that journals from one of Beck's co-explorers on the Whitney South Seas Expedition reported "it was virtually impossible to get a shot off" Beck; "as soon as [the other explorer] would see a bird take off and start to lift his gun, he would hear a crack and watch the bird fall because Rollo had shot it." Dumbacher said that they used a very fine "dust" shot for the smaller birds. Having evolved on the Galápagos without predators, birds did not flee upon approach and it was possible to just walk up to them and pick them up, "so they probably saved shot where possible." Beck was fast with everything, and it took him five to ten minutes to prepare a bird with notes, tag, and associated materials, which generally takes today's collectors an hour. His skinning and stuffing proficiency were such that upon hearing Beck was coming to visit, scientists at the American Museum of Natural History in New York provided dead starlings from the Hudson River and asked for a "Rollo Beck demonstration."

The academy's 1905–06 expedition was motivated to collect evidence of evolution, and the giant tortoises that embodied the concept were on the verge of being hunted to extinction. As Matt James, an expert on the expedition, has described it: "The academy decides to get them before it's too late. Thinking they're better dead and preserved in a museum than left to the whims of fate. They go there to simply harvest the last ones. So when Beck goes to the island of Fernandina he hikes up to the top of this island. Which by all accounts is an almost impossible hike.... He stays overnight, sleeps on lava, he has two hundred ticks crawling all over him.... Eventually he finds this tortoise as the sun is going down so he puts down his pack and eats his dinner with the tortoise munching on grass nearby." Beck noted his intention to skin the tortoise by moonlight, and James opined that this would have taken him most of the night. "Then he puts it on his back, this thing has to weigh like 150 pounds.... He carries this all the way down to the coast on a walk that will kill the average person." James noted that Beck "was just tougher than most people. He required less sleep, less food, less water. He drove people to their graves."[20]

Without a doubt, Beck led the academy's expedition valiantly. It cannot have been terribly easy to keep a group of young men confined on a small schooner happy and motivated for a year and a half. And there were conflicts. After several previous dustups, Beck received a formal letter signed by his eight crew members: "Sir:– Mr. J. J. Parker, navigator of the schooner *Academy*, having proven himself, in our belief, incompetent, the last and most serious exhibition of incompetency having occurred today, (August 9, 1906). We, the undersigned members of the expedition, ask you, the master of the aforesaid schooner, to take the navigation of the said schooner out of his hands."[21]

The next day, the beset navigator replied, in a similarly formal letter. Concurring with the crew's assessment of his performance, "because of the Vessel missing Stays and taking the Ground on Lava Rocks from which position of danger I with the help of the entire crew freed her in about two and one halve hours," Parker concluded that Beck would do best "to agree with the request." Parker offered to continue on board doing "common ordinary labor."[22] This Beck agreed to, though eventually Parker was asked to disembark permanently. Beck assumed the navigation of the *Academy* for the duration of its journey.

When life's big little hiccup rearranged the Bay Area, Beck dutifully fulfilled his day's entry in the ship's log, April 30, 1906: "Learned of earthquake in San Francisco [that] destroyed 5000 people & 300,000 without homes or food. Famine inevitable according to dispatches. Boys want to start for Guayaquil to cable about folks but finally decide to go to Chatham first." One can imagine what the young men aboard were feeling during this time, yet nearly all their subsequent logs' entries document the quotidian sightings and activities of a scientific expedition at the turn of the century, and hardly mention what they must have feared was going on back home in San Francisco. "Skinned birds till noon"; "No land in sight all day." The *Academy* did not in fact go on to Chatham Island but instead left Albemarle eventually to reach Charles Island on May 15, where a load of mail gave the crew its next download about the huge disaster that had struck the expedition's point of origin.

Loomis kept them updated, and told them to delay their return. The various families and friends of the academy's crew were physically safe, but not so the institution that had sent them on the voyage. The quake itself, of course, dealt a major blow to the foundation of the building, which would in any event soon burn to the ground. When the explorers re-entered San Francisco Bay five months later, they reportedly declared, "We are the academy now!"[23] Indeed. The *Academy* returned with over 75,000 specimens, including 264 tortoises, and more than 8,500 birds.

From Slevin's log, November 29, 1906:

Razors were broke out and we all took a shave, some for the first and some for the second time in 17 months. Old clothes are being thrown overboard so the Board of Health will not hold us up, and we are washing in fresh water. At noon we set our course for the lightship and made up to it at 3:00 P.M. Here we picked up our pilot, Captain George Kortz, famous to all mariners entering the port of San Francisco, and shaped our course for the Golden Gate.... With the help of light breezes and a strong incoming tide, we drifted through the Golden Gate and narrowly missed colliding with the pilot boat *Lady Mine*, which was in the same predicament as ourselves. By 9:00 P.M. we reached Lime Point and came so close to the rocks that we put out the ship's boat and tried to pull the schooner's head around. Not succeeding in this, we hailed a passing crab fisherman in his fishing smack *Louisa*, and for the sum of ten dollars he agreed to tow us across to the quarantine station, where we arrived at 10:15 P.M., 65 days out from Culpepper Island and the 519th day of the voyage.

When the *Academy* returned with its cache, the specimens were stored in a walled-off section of the otherwise destroyed building. Not ideal conditions for studying them. At the same time, the decimation of the academy incited even more fervor to rebuild its reputation, and curators began publishing papers on their findings. The flood was initiated in 1907 with a paper on the giant land tortoises; by 1911 there were papers on butterflies and a 280-page botanical survey by Alban Stewart; in 1912, a monograph on snakes. In 1913 Edward Winslow Gifford produced a paper on the birds of the Galápagos. These publications, disseminated to scientific institutions around the world, essentially re-established the museum. Reverberations from personality differences began to be felt; Alban Stewart, a.k.a. "Granny," for example, appears to have coded his botanical specimens in such a way that no one else could ever make sense of them. Rebuffed by the academy, where he was refused a paying job, Stewart went on to a prominent career elsewhere. The specimens he collected for the academy remained in California; later, the botanist Alice Eastwood, charged with figuring out his gnomic categorizing, threw up her hands in disgust and called his work a "Rosetta stone." This is pretty bad behavior for a scientist—to obscure his own research so that others may not benefit from it. Yet, of course, history may eventually find another ending for Stewart's story, if some once or future botanist can figure out his key.

Find Your Niche

Rollo Beck continued to work for the California Academy of Sciences for several years after the 1905-06 expedition, but denied a raise (purportedly due to lack of funds), he jumped ship to join Joseph Grinnell at the University of California's new Museum of Vertebrate Zoology. Grinnell is a pivotal figure in ecology, among other things extending the basic ecological premise of life zones, which tracks the distribution of plants and animal communities according to elevation. Grinnell included precipitation and vegetation among the numbered conditions driving where species live and in what amounts. He broadened the definition of a *niche*, which describes the ecological conditions under which a species lives, formulated originally according to predation, or "who eats you and who you eat." Grinnell saw habitat and behavioral patterns as intertwined, and traceable along an evolutionary trajectory. His deeper definition of *niche* evokes the root of the word that means "nest," and indicates a sense of where a species finds home in the world. He collected birds to establish species boundaries according to geography and niche.[24]

"Environments are forever changing," wrote Grinnell, "slowly in units of recent time, perhaps. Yet with relative rapidity they circulate about over the surface of the earth, and the species occupying them are thrust or pushed about, herded as it were, hither and thither. If a given environment be changed suddenly its more specialized occupants disappear—species become extinct. . . . The course of organic evolution has been molded and is being molded by environmental circumstance."[25] These are key insights into why and how climate change is pulling the proverbial rug out from underneath species, changing their habitats faster than many of them can adapt.

Although he did not see anthropogenic climate change coming down the pike, Grinnell was prescient about habitat loss. Observing the pace of industrial expansion in his own time, he predicted that the flora and fauna of California as it then existed would soon no longer. From 1914 to 1920 Grinnell led expeditions to survey a broad elevational transect in Yosemite National Park. Grinnell designed it for the future, and the results—largely owing to his comprehensive and laser-focused field notes—are one of a kind. Grinnell correctly envisioned that the value of his collections on this project would be made manifest after "the lapse of many years, possibly a century, assuming that our material is safely preserved. . . . The student of the future will have access to the original record of faunal conditions in California and the west wherever we now work. He will know the proportional constituency of our faunae by species, the relative numbers of each species, and the extent of the ranges of species as they exist today."[26] What is colloquially known as the "Grinnell re-survey" has been going on since 2006, with Berkeley researchers consulting more than 3,000 pages of field notes, 4,354 specimens, and more than 700 photographs taken by Grinnell's teams as a baseline for evaluating how species are responding to climate change. You guessed it, they're moving.[27] What was once their niche is no longer.

Beck the swashbuckler and control-freak Grinnell make a funny pair. A significant dimension of Grinnell's legacy is his fastidious, obsessively complete field notebook (you can read it online at ecoreader.berkeley.edu). On a November 1910 collecting trip at Pacific Grove in Monterey, Grinnell notes birds "shot by Beck," and then takes four pages to enumerate "Beck's modus operandi." You can practically learn to skin a bird by reading the description: "Beck uses *main strength* combined with rapid motions and an elimination of all useless movements. Where I would operate with caution and gently . . . Beck exerts all his strength and achieves rapid results." Incisions, yanking, scraping, and sewing are detailed. The next day Grinnell records hunting for birds, Beck ahead of the team, "up on the pilot house most of the time scanning the ocean." Beck comments that dark-bodied shearwaters they sight are due to leave shortly, "those here yet being mostly stragglers as it is," but black-vented shearwaters "are here for the winter," and their numbers are "due to increase." Beck does his chumming, and "desirable species" of birds take the bait.

Grinnell clearly respected Beck's natural history knowledge but as bedevils the amateur, Beck did not have the gravitas necessary for Grinnell to quite take him seriously in the matter of species discernment. Beck went to Grinnell in the 1930s with the evidence of a "few hundred dowitchers"—long-billed wading birds—and asked him to reconsider his "lumping" of what Beck was sure were separate species into one. Grinnell "looked into the issue" but didn't change his mind. "What Beck had observed and others had not," write John Dumbacher and Barbara West, "was that the short-billed species occurred primarily in a salt-water environment while the long-billed species occurred primarily in freshwater environments."[28] It was not until the

1980s that Beck would be fully vindicated by genetic analysis: The long- and the short-billed dowitchers are distinct species.

We now think of both the tortoises and the finches as the starring specimens of Darwin's Galápagos expedition and so they are, but the birds did not get their name in lights until a British ornithologist, working intensively at the academy on the finches collected by Beck, dubbed them "Darwin's finches." Today the finches are the poster birds for explaining natural selection, particularly the concept of adaptive radiation. The thirteen species of finch on the islands have beaks of different size and shape. Darwin didn't particularly notice this, and the birds he mentioned in working out his theory were two evident species of "mocking-thrush" on the islands. David Lack spent five months in the Galápagos studying the finches and their beaks in the late 1930s. Feeling that the birds he'd collected weren't going to do so well on the long trip back to England, he stopped off at the academy to deposit them in the collections and then spent five more months there studying Beck's birds. (He went on to New York and added the sizeable collection at the American Museum of Natural History to his task.) By 1940 Lack had finished a paper on what he'd discerned, "The Galápagos Finches (Geospizinae): A Study in Variation," which the academy published but not until 1945, due to wartime paper shortages. In this monograph Lack proposes that the finch beaks are an "isolating mechanism" that keep species separate from one another. In biologist Ernst Mayr's term, the different bill size is a "species recognition signal."[29]

But then Lack cogitated on his own research and revised his ideas. He published *Darwin's Finches* in 1947, and in this book he laid out a stepwise path by which subsequent finch populations adapted their beak size and shape to available food. At last here was a concrete history in which to see Darwin's theory of evolution by natural selection in action. As Mayr writes, the process of speciation "requires also the acquisition of adaptations that permit co-existence with potential competitors." Establishing that morphological change is related to environmental conditions, he brought the conversation about species back to the world of climate, geography, phenology, and so on—all the physical goings-on in which species abide. Mayr said Lack "more than anyone else deserves credit for reviving an interest in the ecological significance of species." Today, science knows well that species are not just riders on the system of nature, they *are* the system of nature. Yet the fundamentals of nature's story are as yet not completely known. What are all the species, where are all the species, why are they where we find them, and where are they going? How are we going to grapple with all this?

I, Naturalist

On a dry winter day in early 2011 at Stanford University's Jasper Ridge Biological Preserve, the habit of many of nature's denizens to burrow, hibernate, or indeed migrate elsewhere for the season gave momentary pause to Scott Loarie, who was leading a group of docents outside the research station in search of wildlife. Looking around the quiet and seemingly empty woods, Loarie upended a rock. Waiting for the crowd to gather, Loarie herded three revealed beetles with finger prods.

"You could pop one of those in your mouth," someone said, invoking a famous story about Darwin.[30]

"But I don't have to," said Loarie and, as he was debuting the Android app of iNaturalist, captured his beetles by taking their photographs with a phone. The app associates pictures with a GPS reading and the time and sends these records on their digital way to becoming research-grade specimens. The beetles go back under the rock.

Jasper Ridge is a historic place in the annals of conservation biology—the site where Paul Ehrlich and Peter Raven quantified what Darwin had intuited about how species develop traits in tandem with each other, or coevolution (to be discussed a bit later). The sad irony of Jasper Ridge is that as Ehrlich in particular studied a population of Edith's checkerspot butterfly there, it went extinct under his nose. It is a meaningful location in which to debut the crowdsourcing capacity of today's technology to help stem extinction. Loarie was accompanied at Jasper Ridge by Ken-ichi Ueda. With two fellow graduate students at UC Berkeley, Ueda developed iNaturalist in fulfillment of their master's theses in information sciences.[31] Both Loarie and Ueda are handsome young men in their midthirties. In almost comical contrast to the ebullient, charging-forth Loarie, Ueda is quiet and can be cranky. One well-seasoned San Francisco naturalist calls them "Master Loarie and the Prince of Darkness." Ueda has a stern demeanor but in conversation it quickly softens and, in fact, he's a bookish sort of sweetheart. At the time of this outing in 2011, Ueda was working as an engineer at Goodreads, the social networking site for book lovers, to which he contributes his own reviews regularly. Loarie was a postdoc at the Carnegie Institution for Science in Palo Alto, and had racked up impressive papers in journals like *Nature*. One of Loarie's persistent focuses is on predicting extinction using computer models. The trouble with these, as mentioned earlier, is that we do not have many robust data sets to plug into the models. We don't actually have good information about who lives where in what amounts—the baseline for measuring whether species are disappearing or not.

The solution to figuring out how many of any species live where is to count them. Now, that might sound impossible, or at least overwhelming, but not if you are Scott Loarie. And he would seem to have been hatched with giant ambitions. Dr. David Ackerly, a mentor of Loarie's and professor of integrative biology at UC Berkeley, told me about how Loarie's eagerness to take on gargantuan tasks developed into an important paper about California plant life and climate change. "I was teaching at Stanford," Ackerly told me. "And Scott was an undergraduate in one of my classes. We were working on modeling how California endemics will be affected by climate change." The Golden State is, as aforementioned, as distinct an entity as our cultural stereotypes suggest. One might be tempted to say the plant life here exists in its "own private Idaho," except why not just say it makes up the California Floristic Province? Nearly half of our plants are endemic, meaning they are from here and don't naturally live anywhere else. "We were looking at modeling a representative sample of plants," Ackerly told me, "and Scott said, 'Let's model *all* of them.'"

The research took long enough that over the course of it, Loarie graduated from Stanford and headed off to graduate school at Duke University Nicholas School of the Environment. "He kept the study going," Ackerly chuckled, "even from North Carolina." Poring over sixteen plant collections from around the state, Loarie and his collaborators were able to assess the climatic ranges of more than two thousand plants. That's a spreadsheet I would rather not see. The resulting paper, "Climate Change and the Future of California's Endemic Flora," projects that California plants are on track to lose more than 80 percent of their geographic range by the end of the century.[32] The analysis made the newspapers.[33] Loarie had not quite gotten his PhD.

One qualification the paper makes is this: "Projecting the impacts of climate change to an entire endemic flora is complicated by scarce and variable distribution data." How much better if there were abundant and consistent data on species?

When looking at climate change impacts, the consequences of a changing landscape are very high. California, for example, depends on agriculture and viticulture to provide some of our most robust economies. But plants that live here now may not be able to live here in the future—which changes our picture of what California is, and the very industries the state is built on.

"We're at this amazing moment in time," Loarie told me, "with this extinction going on. But here we are near Silicon Valley, and the very same kind of technology that keeps you in touch with your friends through social networking can be used to document species." Before Loarie got involved with it, iNaturalist was entirely Web based, so users would take a photo of a species with a camera, and with a GPS device associate a latitude and a longitude to the image, along with the time it was taken, and then upload the information to the Web site. Latitude and longitude, of course, together create a precise point in space, so iNaturalist records exactly where and exactly when a species is observed. "With iNaturalist," Loarie said, "you can confirm that species are living in the places we think they're living. And you can start to see where species are being observed outside their historic ranges." To the seemingly outsized task of documenting all biodiversity, iNaturalist has the solution: crowdsource it.

Joining forces with Ueda, Loarie quickly mobilized a Global Amphibian Bioblitz. The origins of the bioblitz are discussed later in the narrative, but suffice it to say the term usually refers to a one-day rapid inventory of a discrete area, often a park. Most bioblitzes enjoin volunteers to go out and make observations of every taxon they can find, from bird to reptile to bug, but as with the amphibian project, they can also be used in a more focused way. As hundreds of users almost overnight joined the ranks of iNaturalist, projects using the tool today include Global Bat Watch, State Flowers of the United States, and Redwood Watch.

The Android app has been joined by one for iPhone. Users routinely scrutinize uploaded observations and the crowd self-corrects the species designation. When there is agreement that a species is, yes, that species, the record gets uploaded to the Global Biodiversity Information Facility (GBIF), which as the name implies is a universally accessible biodiversity database in use by scientists all over the world. For example, a new species of frog entered the scientific lexicon through iNaturalist when a herpetologist in New York City viewed a photo taken in South America and flew down to investigate his hunch. Today the number of iNaturalist users grows daily—at last tally, it logs one observation of nature a minute, and has surpassed the two-million-observations mark. "We're nowhere near big enough," Loarie said, citing the hundreds of millions of observations garnered by eBird, where amateur and professional birders document sightings from all over the world. "eBird is real science. We want to do for the rest of taxa what eBird has done for birds. Crowdsourcing has the potential to scale biodiversity data on the model of Facebook and Google."

In 2011, Loarie and Ueda significantly ramped up the reach of iNaturalist, bringing whole countries to the platform (including Costa Rica, New Zealand, and Mexico, which have all contracted with iNaturalist to provide a template with which they can inventory their biodiversity), but they had a problem: They weren't being paid by anyone. Loarie was tarrying at Carnegie, but you can't be a postdoc forever. One peer of Loarie's said to me, "He has to get a real job." Loarie resisted this because he believed in iNaturalist, he wanted to follow his bliss on iNaturalist, but realities were involved. Conventional wisdom would find Loarie, so gifted and successful in academia, practically crazy to turn away from it. A friend from Carnegie got big funding to develop technology to help agriculture figure out how to deal with climate change, and

Loarie considered taking a job with him. "All the modeling skills are the same," he told me, whether you are analyzing stocks on Wall Street, percentage of ground cover in the tropics, or numbers of species and where they are distributed. Loarie held out, sustaining his conviction that developing iNaturalist was the right thing for him to do, with or without institutional support.

In the spring of 2013 I sat in on Loarie's class in natural history taught through the lens of iNaturalist in the Department of Geography at UC Berkeley. Geography at Berkeley is mostly geared to human populations, so most of the students were not versed in science per se. "Climate and geography will come up again and again," he told the students, "but we are really talking about plants and animals—living things." Over the semester he taught the basics of the *scala naturae*, the great chain of being, a concept handed down from antiquity that provides natural history with its original template, and explained Linnaeus and the binomial naming system. He explained life zones and laid out the question of scale, from species to community to ecoregion to continent to globe. He showed NASA satellite photos and said, "No matter how good the satellite data is, we aren't going to get individual species. We need to anchor all this to the ground." When you make an observation on iNaturalist, then "we can say that these species are living here and interacting with their environment." Elucidating major ecological concepts one day, he sketched the species-area curve on the blackboard and commenced to talk about it. Then he stopped. He was silent. This was most unlike Scott Loarie. A bright voice from the back of the class piped up with a concise explanation that sounded right to me. Loarie apologized and dismissed the class a few minutes early.

He sat across from me at a long table and put his head down. I was slightly alarmed at the sight of his black hair flopping over his arms. Was he sick? He sat up. "I bonked," he said. "I stayed up all night finishing a project for GBIF." The thing is, the guy did not actually look tired. "I guess I have limits! I need *some* sleep." At the next class he again apologized and gave a thorough explanation of the theory of island biogeography.

One day after class I went with Loarie to meet Ueda for coffee and listened in on one of their powwows. The academy had tacitly presented itself as a potential home for iNaturalist, but promised funding for a big citizen science push at the institution had evaporated and there was no clear indication of how they might go forward with it. Loarie paid Ueda a salary from contracts to build custom iNaturalist projects. These were small sums of money and while every project amplified iNaturalist in the world, building little nodes in an ad hoc way wasn't great for the overall functionality of the site. Ueda got particularly cranky about what he called "feature creep," though as Loarie noted, he went to great lengths to meet the needs of individual users.

"No matter what happens, we will keep iNat going," said Loarie. "Even if it grows slowly and incrementally, we will keep it going."

Ueda had joined us via his bicycle and was wearing a knitted cover on his helmet in the shape of a nudibranch. He seemed to take their uncertain future in stride.

"I'm more interested in the citizen naturalist," Ueda told me, when I asked him about the potential for his invention to connect people with scientists to help stem extinction. "I started iNaturalist as a lifestyle project, as a way to help people get more attuned to the natural world on a day-to-day basis; to record and share usable data not because they are contributing to a cause but because it's fun." He explained: "I'm interested in identity, how we define ourselves, and how we connect with each other. I don't know if iNaturalist has ever converted anyone who wasn't already inclined to be interested in species distribution. But it has definitely helped people become attuned to things outside their original interests. Someone will write, 'I found this rare flower and I knew what it was because I saw it on iNat,' and that's what makes me really happy. You see lichens on iNat and you become attuned to lichens. The mission of iNaturalist is to connect people to nature through technology. Personally, I care most about that single comment from someone who said, 'Now I know that species!'"

In 2014 iNaturalist became the official mechanism for making species observations for the National Park Service bioblitzes. At the conclusion of its Bay Area bioblitz in 2014, National Geographic's vice president for research, conservation, and exploration, John Francis, declared, "iNaturalist has changed the way we do bioblitzes."

The Disappeared

Perhaps the most profound way we are altering earth processes and heading toward a tipping point is by way of species extinctions. Although we are all used to responding to desperate pleas for polar bears, elephants, and tigers—among other threatened megafauna—extinction forces are bearing down on species all over the globe at all different trophic levels. It can be difficult to see where less-visible extinctions are occurring, and iNaturalist can help. This is the unexpected counsel of Stuart Pimm, professor at the Nicholas School of the Environment at Duke. Pimm is an expert on extinction, and he points out that our current lack of information about where species are and how their populations are faring is a big impediment to rescuing them.[34] iNaturalist could help better focus our protection efforts.

The fundamental science of taxonomy involves naming species and putting them on the tree of life. Pimm says this practice is more important than ever, and he suggests the rest of us help the discipline along by using iNaturalist. Pimm is in the grandfather generation of conservation biologists, and it is unusual for these senior guys to step across the traditional line dividing pure and applied science. But Pimm really wants to save biodiversity, and he's eager to purpose technology and the crowd to help. He suggests that this citizen science tool available for wide use by people with and without PhDs could help create "a public species' range map." This would be updated, amended, and confirmed by users. Pimm suggests this local data be combined with satellite imagery at larger scales to create "metapopulation models of fragmented ranges." These could zero in on where species are losing habitat. The idea is to have a live-streaming, constant monitoring operation tracking biodiversity. We could respond quickly when we see things going downhill for a species or communities of them.

The idea of a map of the world has been both a major tool for exploration and a canvas for humanity's dreams about itself for perhaps thousands of years. To create a literal inventory of the world is to tame our inevitable distortions about what is actually where. Since Noah's Ark we've wondered where animals and plants have come from, when and how they got there. With an iNaturalist map of the world's species, we could track their heroic journeys. Bioblitzes to gather data can turn up species nobody knew were living in, say, a municipal park. By the same token, iNaturalist can be an early warning system for detecting invasive species before they get real purchase on an ecosystem. In 2015, Rebecca Johnson from the California Academy of Sciences organized a bioblitz at a citizen science conference in San Jose, California. Scott Loarie led a group out to the patchy natural areas at the edges of the giant convention center. Relative to other bioblitzes, this one was brief, because everybody had to get back to the conference. A mom using iNaturalist with her kids photographed a snail that turned out to be an unwelcome newcomer to Silicon Valley.

When people talk "real" science they often invoke hypothesis testing, which is generally outside the realm of inventory and monitoring. First you have to define a question, create parameters around what kinds of observations you will make, or establish a context for those observations. Some basic questions can be framed as hypotheses. For example, "what animals are likely to be hit by vehicular traffic when and where" could establish a project seeking to minimize roadkill. To make their data usable, drivers would document a start time, finish time, and distance traveled to ground their observations of dead animals.

Scaling It Up

While iNaturalist is accumulating species observations in the millions, Loarie's ambition is for it to accumulate hundreds of millions of observations—to get to the point where global patterns can be discerned in the accretion of pixels that together make up a picture of life on earth. The kind of science that looks at these patterns is pretty academic, and it's definitely big picture. But iNaturalist is nimble and can be applied at more local scales to great effect.

Take our favorite disaster for wildlife, the oil spill. On May 19, 2015, a ruptured pipeline belonging to the Plains All American Pipeline disgorged 140,000 gallons of crude oil in the vicinity of Refugio State Beach in Santa Barbara County.[35] "Paul Hobi from the Marine Protected Area Collaborative Network was worried about communicating damages and impact quickly," Rebecca Johnson told me. "He had made some KML files of the extent of the oil sheen." *KML* stands for "Keyhole Markup Language," and the discerning reader will quickly realize its connection to what evolved into Google Earth. KML is essentially a file format that displays geographical information on your computer, in the form of a polygon on a map. "I was curious about what we knew about this place," Johnson told me. So she "created" the place on iNaturalist, defined by the boundaries Hobi had determined, and she pulled out all the species observations that had been made in that area and posted on iNaturalist. Largely because a class from the University of California at Santa Barbara had inventoried the area the weekend before the spill, there were about three hundred observations.

Johnson plays a unique role at the California Academy of Sciences. On the one hand she's a traditionally trained invertebrate zoologist. In this world, if you find a nudibranch no one has seen before in the place where you find it, that nudibranch becomes the subject of a scientific paper. Questions about why the nudibranch is "out of range" are posed in an evolutionary framework—possibly the nudibranch is altering some element of its physiology to adapt to a new environment. Very interesting. On the other hand, she has been highly involved with the Marine Protected Area Collaborative, and the Greater Farallones Marine Sanctuary in particular, for years. She has been closely involved with management decision making about how and where to best protect the rocky intertidal environment. She bridges academic and applied science.

"When people use iNaturalist and share what they are seeing, we can get all this information about a place and we can get it almost instantly, just using the Web site. We don't need to go to Pete Raimondi and ask him for his data. We don't have to set out transects or quadrats. It's not population data, it's not the same as a complete biodiversity inventory or drill-down academic scrutiny, which are critical. But it gives us a pretty good idea of what's where and when, based on reasonable looking. And it's available immediately, which is key for oil spills."

The ability to so quickly marshal relevant information about the Refugio spill inspired Johnson. There is a big thought balloon over the heads of many practitioners and opinionators training their noggins on citizen science. One-off events like bioblitzes do a lot. They engage people to connect to a place and to each other, and encourage them to really see what's out there often quite close to home or somewhere easily accessible. They tune people in to nature. They help accumulate data points that are of incomparable relevance, since they capture a moment in time that will never come again. But in terms of really saving nature, bioblitzes are not enough.

The coastal monitoring world on the West Coast is tantalizingly poised to bump individual and local efforts up into collective impact. Thanks to established ocean-monitoring programs, the nodes of a comprehensive network keeping tabs on the delicate and vital intersection between land and sea are largely in place. The Marine Protected Area network, supported by California foundations and mandated by the California Marine Life Protection Act, includes educators, nonprofits, research institutions, local and regional law enforcement, the California Department of Fish and Wildlife, and in some places, direct municipal involvement in the marine protected areas. All these folks could be asking more synthetic questions about the coast together, communicating more effectively to the public. Money for this, of course, is lacking.

Johnson hit on an idea that makes both a big step in the efficacy of the network and taps iNaturalist's potential to truly monitor nature in place over time. "I thought, let's document biodiversity on our coast," she told me. "Let's get a big picture of what's going on over a broad geographical area, not just one place where you do one bioblitz, but bioblitzes across a big geography," she said. "This kind of endeavor is only possible because of the existing, well-supported network, which I've been a part of for years." With significant help from the Marine Protected Area Collaborative Network and the Resources Legacy Fund Foundation, Johnson and Alison Young at the academy organized Snapshot Cal Coast. More than twenty bioblitzes along the coast were scheduled to take place June 1 through 12, 2016.

In addition to its breadth, Johnson's concept for Snapshot Cal Coast has a depth bioblitzes usually lack. "This is the first time we've gone out at this kind of scale with specific questions in mind," Johnson told me. She compiled a "ten most wanted" species list to direct observers on their hunts. These include sea stars, with particular attention on whether they show signs of wasting; seaweeds, including those that are in decline at various locations; and nudibranchs. Hopkin's rose is on the list—with Johnson's data from the week, a picture may very well emerge that helps us understand better why and how we are finding it in such high numbers in Northern California. Snapshot Cal Coast will also pay particular attention to invasive species—for example, spiny lobsters that are moving north.

In her dual roles, Johnson often bumps up against traditional science protocols. "We've been trained that you keep your data private until you publish it," she told me. "If a volunteer found a new nudibranch in San Mateo, they would contact you, the scientist, and that was a paper. Now it's an iNat observation, with a community conversation, a tweet, and a blog. As a scientist, I still have something important to contribute, but every discovery doesn't have to be filtered through me and my inbox." Data has a direct line to the public. As an ambassador of citizen science, Johnson shines. "I find joy in it," she told me. "Deputizing and empowering people is the best part of the work. The community is everything—together we are learning better to discern place, to love our place, and to function as an early warning system to help better protect it."

Terry Gosliner was a consistent booster for iNaturalist at the academy, and in mid-2014 a contract was signed. The arrival of Jon Foley was good news for Loarie and Ueda, since he naturally understands and appreciates the use and power of big data sets and social networking—it is not forgone that someone at the head of a natural history museum would have that appreciation. "The rate of collections can't keep up with extinction so we have to do something else," Foley told me. He's excited about the observations collected by iNat but also its potential "as a platform for researchers all over the world." Foley envisions iNaturalist providing a lingua franca that will allow conversation and cross-pollination to occur across all scales of ecological inquiry and measurement, from government and university researchers on the one side to kids doing bioblitzes on the other, and everything in between. "This is where technology really allows us to scale," he said.

More Things in Heaven and Earth
Periodically I brave the monthly "Philosophical Pizza Munch" run by Mike Ghiselin and Elihu Gerson at the academy for about twenty years now, in which a shifting group of academics and researchers from a panoply of disciplines debate the nuances, virtues, and flaws of selected peer-reviewed scientific publications. I am curious about the "philosophy of science," wondering if it is the very antithesis of citizen science. Philosophy must be the purest of the pure. This is Mike Ghiselin's thing, and I want to understand better where he is coming from in objecting to applied science. Lo and behold I find that Gerson, who is an

independent sociologist, and Jim Griesemer, a professor of philosophy at the University of California, Davis, are the authors of several significant papers, focusing on the case study of Joseph Grinnell at Berkeley, to discern just what respective roles in science the amateur and the professional should play. Roberta Millstein, also a munch regular and also a philosophy professor at Davis, has a special focus on environmental ethics, a field that intersects with both the philosophy of science and the worlds of everyday people. Ideas, it would seem, are often applied.

"Scientific collaboration is usually conceptualized as the work of teams of scientists with shared goals," write Griesemer and Gerson in "Collaboration in the Museum of Vertebrate Zoology," including formulating and testing hypotheses.[36] The purpose of this paper, the authors state, is "to suggest that we need a wide view of collaboration," because the work of research involves more than scientists. The story of Annie Alexander, heiress to the C&H Sugar fortune and benefactress exemplar to the University of California's Museum of Vertebrate Zoology and Museum of Paleontology, both of which she founded, is fully told in Barbara Stein's biography, the title of which pretty much states how Alexander accomplished all this: *On Her Own Terms*.

What particularly catches Griesemer's and Gerson's attention is not the fact that Alexander placed Grinnell as head of the new museum and bankrolled him, as philanthropists do, but that she fundamentally helped shape, drive, and fulfill its research agenda. In her early thirties, Alexander started attending John C. Merriam's paleontology lectures at Berkeley, and she was hooked. "I like it more and more, this study of our old, old world and the creatures to whom it belonged in the ages past, just as much as it does to us today. Perhaps the study is all the more interesting because it is incomplete."[37] Alexander went on fossil-collecting expeditions, and with her father on a safari to

Africa in 1904, where they followed the form of the day and shot large carnivores to take home. At around this time Alexander became aware that as fossils tell the backstory, living animals tell the earth history narrative as it is unfolding. Working with diligence and zeal that Rollo Beck would have appreciated, she set to preparing the skeletons of the African animals herself: "Such a boiling and a stewing that has been going on!"[38]

Annie Alexander was loosened from some of the conventions of her day by way of her wealth, yet she was a woman of her time who did not finish college and, despite her astute comprehension of the project of natural history, felt intellectually inadequate to pursue it professionally. Yet she was no shrinking violet. Over the course of a long and productive life, she held her own with prominent men who were, yes, eager to benefit from her financial largesse, but who would perhaps have also liked to control her purse strings to more specifically suit their personal objectives. Alexander frequently held to a different view. She was in her thirties when her father died on that 1904 safari; crushed, she later explained, "I felt I had to do something to divert my mind and absorb my interest and the idea of making collections of west coast fauna as a nucleus for study gradually took shape in my mind."[39]

By 1906 she was hatching plans for a natural history museum to house her specimen collections. C. Hart Merriam, then head of the Biological Survey in Washington, DC, encouraged her to start a new museum rather than invest in "existing museums on the Pacific Coast," commenting that these, "as you so well realize, [are] only feeble attempts along certain more or less ill-defined lines."[40] His dis to the academy reflects its down-for-the-count status upon the 1906 earthquake, and also points to the eventual utility the Galápagos specimens would provide. They would establish the academy as a serious taxonomic research operation in the largest context of its application—the discernment of evolution by way of natural selection—based on the very creatures that gave Darwin his idea to start with. For her part, Alexander felt the academy wasn't sufficiently interested in mammals and that her collections would molder in the basement. She set about looking for a scientist to work with and ran into Joseph Grinnell at the Throop Institute (now the California Institute of Technology). They hit it off.

Grinnell was vision-ready with his concept of studying evolution from the perspective of biogeography. As Griesemer and Gerson point out, "Grinnell's museum was to be a sensitive instrument for measuring evolution. The time scale of evolution is such that the museum instrument would need to operate far beyond Grinnell's own career and lifetime." And so it has. Alexander enthusiastically helped fill the museum with specimens, collecting them according to Grinnell's strict guidelines for including detailed information about where, when, and what else was going on when each specimen was taken. He was no fool and consistently deferred to Alexander and respected her opinions (others who wanted her money were not so smart and tried to override her). "The Museum was designed as an institution for conducting 'big' science," say Griesemer and Gerson. And such a thing cannot go on without collaboration between amateurs and professionals.

Satisfied that two philosophers understand that science occurs in a context that is composed significantly of nonscientists, I was even further delighted to read Roberta Millstein's "Environmental Ethics" contribution to the textbook *The Philosophy of Biology: A Companion for Educators*.[41] Millstein notes in her abstract that "students . . . come to me with pent-up questions and a feeling that more is needed to fully engage in their subjects" concerning the natural environment, how we value it, and how we ought to behave toward it. Who or what is part of the moral community to which we owe direct duties? "Is it humans only? Or does it include all sentient life? Or all life? Or ecosystems considered holistically?" Facing these and other questions of definition "will make students more reflective and thoughtful citizens and biologists, sensitive to the implications that different conceptual choices make."

Millstein starts with Immanuel Kant's categorical imperative: "Act in such a way that you treat humanity, whether in your own person or in that of another, always at the same time as an end and never merely as a means." She goes on to Jeremy Bentham, who questions the "insuperable line" between humans and other animals: "The question is not, Can they *reason*? nor, Can they *talk*? but, Can they *suffer*?" And Peter Singer's principle of equal consideration of interests, upon which she muses that this poses a problem when one needs to kill plants to eat them. "All biocentrists acknowledge this, and all have developed ways of trying to balance competing interests but the challenge is to do so without continually defaulting to the interests of humans or by developing an ethic that humans are unable to live by."

Millstein lands all her balls squarely in the court of Aldo Leopold, who one might imagine is posthumously surprised at being invoked so powerfully by a philosopher. Born in 1887, Leopold was a game warden and eventually the author of the first textbook on wildlife management. He was a prolific writer, but his classic *A Sand County Almanac* took years of revision before being accepted for publication in 1949. (*The Hero with a Thousand Faces* underwent a similar process and was published the same year.) The *Almanac* is a deceptively simple meditation on the yearly turn of seasons, migrations, and so forth, and in these subsequent sixty-five years his powerfully understated and indeed philosophical wisdom has slowly descended through several generations who now revere him as the patron saint of wilderness. Leopold's essay "The Land Ethic" appears in the *Almanac*, and here, as Millstein amplifies, is a citizen science manifesto.

In simple language Leopold says that since we humans are interdependent with "soils, waters, plants, and animals," we ought to include all of the above in the community to which we owe direct moral duties. Encapsulating the trophic cascade and connectivity concepts that conservation biology would later quantify, Leopold explains that humans are part of an interdependent system. The land ethic is simple: "A thing is right when it tends to preserve the integrity, stability, and beauty of the biotic community. It is wrong when it tends otherwise."

In a section of the *Almanac* titled "Odyssey," Leopold limns the heroic journey not as undertaken by the human hero of Homer's epic or by any human at all. Instead he chronicles the passage of "an atom locked in a rock," which he calls "X," through its release over the course of geologic time. He describes the recomposition of X through erosion into soil which then becomes a flower, an acorn, a deer, and then an Indian—sequential transformations are achieved by way of the trophic cascade. "From his berth in the Indian's bones, X joined again in chase and flight, feast and famine, hope and fear." Joseph Campbell asserted the heroic journey as the pathway to personal identity. Leopold embeds personal identity into a larger context made up of shape-shifting carbon molecules; when possessing the temporary form of biotic life, "creatures must suck hard, live fast, and die often."[42]

Millstein brings the philosophical and the practical together. "We can seek to answer the question, 'what is biodiversity?' but that question very quickly becomes 'what should we preserve?'" She says that "conceptual debates sometimes occur outside of the normative realm," but once you find yourself in the real world, as it were, "it is hard to keep the normative issues separated from the conceptual ones." The way Leopold puts it works both philosophically and practically. His land ethic calls for changing "the role of *Homo sapiens* from conqueror of the land community to plain member and citizen of it."

Party Pooper

But what is even practical philosophy when measured against actual events? The dark mark of *Homo sapiens* in the geological record, the purposing of more than our fair share of photosynthesis at the expense of other living creatures—these and other accretions have built up steadily despite any and all pretty reasoning that we shouldn't let them happen. A few months before Christmas 2012, my eyes grazed the bookshelf in my living room and lighted upon *Moby-Dick* by Herman Melville. I pulled down the book and opened to the scene where Ahab has assembled the entire company of the *Pequod* to goad them into "raising" the white whale he seeks to destroy. Ahab is so furious at the whale that Starbuck accuses him of blasphemy. But blasphemy is Ahab's pathway to vengeance. "All visible objects, man," cries Ahab, "are but as pasteboard masks." He said, "If man will strike, strike through the mask!"[43]

Melville also provided his own foil or corrective to the cheerful discovery of life Darwin hit upon in the Galápagos. Having visited the islands in his twenties, Melville later penned a black ode to the place in his 1854 work "The Encantadas, or Enchanted Isles." Darwin reports funny interactions with the tortoises, imagining a reptilian stare is asking, "What made you pull my tail?" He tries to ride them, but keeps falling off. Melville by contrast warns that the tortoises have a dark side. "Enjoy the bright, keep it turned up perpetually if you can, but be honest, and don't deny the black." Darwin finds in the Galápagos an astonishing "amount of creative force," while Melville wonders "whether any spot on earth can, in desolateness, furnish a parallel to this group." Melville dreams that three tortoises are holding up the earth as in a Hindu creation story and the next day, he eats the tortoises. Darwin cleaves to the new life part of creation and Melville completes the thought, with death—by eating.

No writer, artist, or thinker has succeeded in stopping the onslaught of Ahab's spiritual progeny. Joseph Campbell sunnily offers us the hero prototype but he doesn't much grapple with the hero's dark side. Campbell said the hero's journey is about "overcoming the dark passions," and "symbolizes our ability to control the irrational savage within us."[44] But Ahab is super-rational. In sum, he wants to subdue and destroy nature. As Melville suggests in "The Encantadas," nature itself—life itself—is the evil to quell.

Ahab is a bad guy but even our good guys are highly suspect. As literary critic Richard Slotkin puts it, "An American hero is the lover of the spirit of the wilderness, and his acts of love and sacred affirmation are acts of violence against that spirit and her avatars."[45] Junípero Serra, saint/devil. Thomas Jefferson, liberator/slave owner. It would seem the real "breakthrough" we need is not an ability to invoke a private transcendent joy but rather a willingness to hear both sides of the communal stories we have been telling ourselves.

I had a very uncomfortable realization. I have read and loved *Moby-Dick*, and it has enlarged my soul, but basically it has done nothing else. What would Melville make of the Anthropocene? He warned us, but his book did not slow it down, not one whit. Despite devastating wars including genocides that go on to this day, the human species has only increased in numbers. We have burst through natural checks; we are fast turning the Garden into a series of delivery stops for Internet-ordered goods; we are taking the life out of life. We are Ahab. What made me very unhappy is that as one who lives by the book I realized that books are worse than ineffectual. Experiencing emotional release upon reading a book to me has felt like an action, but it has not been an action.

Nearly two years before my father's death, and well before any hint of it was on the horizon, my parents and other extended family gathered for the holidays at our house. (It is hard not to regret that at the time I did not focus more on how lucky it was to have everyone together, since it turned out to be the last time this would be so.) On Christmas Day the feast was feasted and we played board games in the living room, and I talked about the utility of the liberal arts education with four college-age cousins. The great thing about smart young people is they are completely up to talking about meaning and value and don't as their parents do change the subject back to kitchen remodels. But my daughter was giving me a pointed look. She pulled me aside and reminded me that I was talking to people who were in the midst of undergraduate careers, so it wasn't perhaps a great idea to tell them that literature is a snare and a delusion.

But but but! I was not saying Melville and all the other great minds they were absorbing are a waste of time, not at all. I tried to make the words coming out of my mouth sound reasonable. The last thing on earth I wanted to do was discourage these earnest pursuits. All those years of straight As and rowing teams and getting into good colleges, all that tuition, all those hopes and dreams in those young hearts and in the hearts of their parents!

My brother Jack, who is a therapist, looked at me with alarm and concern. Tears started into my eyes. I was now fully inhabiting my role as a shrill Ingmar Bergman character, breaking up the party. And what was *Bergman* so worked up about anyway? Bergman hasn't changed anything either! "You are really upset about this," my brother said. "Mass extinction," I croaked. "Aren't *you*?"

I made matters worse the next day by taking some family members, including my parents, to see Lars von Trier's film *Melancholia*. I thought it was one of the best films I had ever seen, as von Trier goes full-on into Earth's obliteration, putatively by force of colliding planets, but symbolically taking on capitalism and its discontents. The implacable film seemed to me to confront exactly the "fear and trembling" and the "sickness unto death" of Kierkegaard. It isn't exactly fear of death that makes the film terrifying—it's the fear of instant extinction. Extinction, I realized, is not just about the death of species. It is about the death of meaning. It is about negation itself. Why, I wondered, don't more people grapple with the terrible erasure of creativity, fecundity, and loss of just plain old *life* that mass extinction poses?

Suffice it to say I was the only person in the movie theater happy to experience this film. As the credits rolled the small audience groaned and grimaced at each other across the aisle. "Well I'm not going to worry about climate change anymore," my father said loudly. Everybody laughed and was hugely relieved, shaking their heads as if to get rid of the film. We filed out, reconfirming with muttered complaints that we were all still here and about to go home to a nice dinner, so fine, thank you very much, Lars von Trier!

There was only one thing for my downward spiral and the next day we piled into the minivan. I was at the wheel. My son and daughter, my father, my brother, and my sister-in-law were aboard. We went down to Año Nuevo and watched the elephant seals. These are huge, seemingly immobile blobs of sass on the beach, until a big male decides to move and then watch out! The males are so awful—they bludgeon each other in sumo-style smackdowns and make each other bleed. The females and young ones get out of the way fast when one of the cranky papas starts to throw his blubber around. The elephant seal is brought to us courtesy of twenty-five million years of evolution. They can dive five thousand feet and stay under water for up to two hours. Hunted for oil, of course, they were down to fewer than one hundred individuals at the turn of the last century. But they have made a comeback. A smaller animal that had been watching me approach with big liquid eyes suddenly opened its hose-like mouth and howled at me. Yes, yes, you are right—I backed off.

CHAPTER EIGHT

First There Is a Mountain

On an overcast day in June 2012, Andrea Williams, vegetation ecologist with the Marin Municipal Water District (MMWD) in Northern California, led me and a handful of volunteers to take the measure of grasses, wildflowers, and shrubs at Rock Spring on Mount Tamalpais. This was the inaugural day of a citizen science project that would go on for three years, led mostly by the MMWD with a significant assist from the California Academy of Sciences. Though we would stay put there for quite some time, Rock Spring is a good place to start a hike. The Cataract Trail will take you on a tour of the mountain's signature features, through Doug fir, tan oak, California bay and madrone forests, through grasslands and along a waterway that during rainy winters hurtles to earn that term *cataract*. Rock Spring became the starting point for a regular meditative circumambulation of Mount Tam, as this hunk of rock and age is colloquially known, initiated by the poet Gary Snyder in the 1970s. Snyder reports following the directive of a Hopi elder who stopped at what is called an "outcropping" of serpentine rock—ancient, greenish, crumbly looking—and said, "This is a special place." The serpentine has to do with why Williams chose this spot as well.

We spent about six hours at Rock Spring, pulling a tape measure here and there; marking distances and individual plants with colored flags; digging up representatives of various species, noting whether there are a couple more of them, scores, hundreds, or even thousands of them around; taking photographs; and pressing the plants between newspaper and cardboard in preparation for mounting later. We had been trained in all this by Williams and other botanists, including Frank Almeda, a longtime curator from the academy. Almeda has a big head of snowy white hair and the fastidiousness characteristic of his discipline. Take more of the plant rather than less, Almeda advised, since you can't glue parts back on. Almeda dug up a native iris and cradled it, the root ball in the crook of his elbow and the flower in his hand, as if he were delivering a baby.

Williams has a thing for grasses, partly "because they are often overlooked." There can be more than forty-five species of grass in a twelve-inch-square patch of Mount Tam. We were not sampling all, but many of them, and they are darn hard to tell apart. Thinking that as a journalist I would have a natural affinity for the job, I took on the role of scribe with another volunteer. We held a spindly stalk against a piece of cardboard and took several pictures of it; these were documented by number on a data sheet along with the species identification, where we found it, and whatever details of the plant or its surrounds we could muster, like "tangled in manzanita." We were on our knees in mud and finding handholds on rocky hummocks. None of this was hard, but it took continuous effort. Many a day that starts wet and cold in the Bay Area winds up hot and sunny by midday, but not today. We were damp and cold, then wet and cold, and finally tired, wet, and cold. Silently I invoked Gary Snyder's "Serpentine

Power Point" for psychic sustenance.[1] Just thinking about the harsh green rocks persisting for millennia spurred me on.

As did the ghost of Alice Eastwood. A curator of botany at the academy from 1890 to 1949 (with no college degree), she made weekly collecting trips on Mount Tam and was pretty much never deterred, much less by a little fog. Gary Snyder sees her from the corner of his eye, "her ragged skirt walking off through the bright blue ceanothus or the little white bells of manzanita flowers." "She had a house on one ridge of the lower mountain for over a decade," he writes. "I think now she was the manifest spirit of the guardian mountain who appeared for a while in human form to help point us toward redemption—the redemption of all species, all systems, as one big organic family."[2]

Eastwood amply fulfills the hero's journey in Joseph Campbell's definition, all the way down to the roots of what he calls the "basic mythological problem" of aligning ourselves with nature. "Move into a landscape. Find the sanctity of that land." This is accomplished by "what's called *land nám* by the people in Iceland; naming and claiming the land through naming the landscape, land-taking."[3] Eastwood's successor as curator of botany, John Thomas Howell, is with her responsible for the naming of the flora of Mount Tam, but today it is Eastwood who exerts charismatic pull. People follow her, retrace her footsteps, and see her on the mountain. And here I have to differ significantly with Snyder in one respect. I see Eastwood on Mount Tam too but *never* in a ragged skirt. Among the virtues Eastwood effortlessly modeled in a long life (she died at ninety-three) was a sartorial formality in the nineteenth-century mode. Eastwood wore big long skirts and pleated blouses buttoned to her chin. She also wore fabulous, flower-strewn hats.

Alice Eastwood is arguably the main reason the academy partnered with the MMWD project as a great entry point into its own effort to develop a citizen science program. The academy has a comprehensive collection of plants from Mount Tam, and most of the specimens were collected by Eastwood, who died in 1949, and Howell, who died in 1994. While there were lots of good reasons to inventory the mountain, the academy had the added incentive that specimens we collected there from 2012 to 2014 would have a meaningful baseline, the point of comparison to the historic specimens collected mostly by these two. Put together it took Eastwood and Howell more than one hundred years to collect a representation of Mount Tam, and we didn't have time to do it quite their way anymore.

The water district itself—a publicly owned utility that stewards a commonly owned natural resource—manifested the citizen side of the citizen-science equation. As the Bay Area in general and Marin County in particular got itself organized in the decades following the gold rush, water was provided to the people of Marin by a handful of private companies, most of them subsidiaries of real estate businesses. In a paroxysm of progressivism, the MMWD established itself as the first of its kind in 1912—owned by the people for the people. The idea was that there should be public control over common natural resources.

Tromping back to the truck that would transport us and our specimens back to the Throckmorton Ridge Fire Station, across Panoramic Highway from Alice Eastwood Group Campground, Williams said the idea for the re-inventory came to her while she was trying to think of a good way to commemorate the district's centennial. In line with her staunch appreciation of grass, so often ignored and even trampled on by the rest of us, Williams was now sticking up for the dirt and rocks it grows in and the water percolating through its roots. She pointed out that the MMWD has done more than bring water to the people; it has also saved the watershed from development. Twenty minutes from San Francisco and you can get on a trail that takes you from bay to ocean. Do it at the right time and the only other souls you will come across have a different species designation from your own.

Of course it is not only the MMWD that deserves credit for protecting all that we enjoy on Mount Tam. State, federal, and local agencies have all done their part to keep the mountain free of development. A lot of the continuing blooming and dying glory here is due to the philanthropy and politics of William Kent, a progressive Republican congressman who among other things helped establish the water district and also made a gift to the people of the most visited 298 acres in the country, by way of the Act for the Preservation of American Antiquities in 1906. President Teddy Roosevelt wanted to name it Kent Monument, but the benefactor demurred. "I have five good husky boys that I am trying to bring up to a knowledge of democracy and to realizing a sense of the 'other fellow,'" Kent responded, adding that his sons should be able to keep his name alive without adding a monument into the mix.[4] Kent's wish was granted that his donation be named Muir Woods. The honoree was pleased. "This is the best tree-lover's monument that could be found in all the forests of the world," said John Muir.

The late afternoon light bathed us in a whitish haze through persistent fog and the landscape was obscured. I was very likely the least hardy person in the truck yet probably not alone in having some second thoughts about where I would lay my head tonight. As this first day of the survey was the official marker of the MMWD centennial, all of us volunteers had been offered a special privilege—to camp out at the Mountain Theater. There are several campsites on Mount Tam, but the Mountain Theater is located in a sweet spot above Rock Spring. The amphitheater hewn into the hillside looks out over San Francisco Bay, at the twinkling lights of Sausalito and any night-cruising boats, to the Golden Gate Bridge and the dark rise of Angel Island. Camping, however, has been forbidden here since the place was overtaken by almost forty thousand Summer of Love sybarites in June 1967, come to groove with the Doors and Jefferson Airplane (transported up the mountain by Hells Angels). Apparently you can have too much fun on the watershed. Not tonight—the cold was getting colder.

"The watershed gives us so much," Williams said, sounding more like Gary Snyder than like a vegetation ecologist. "So I was thinking, what could we give to the watershed?" A watershed is not actually water but earth: the rocks, soil, vegetation, and animals through which water cycles. Try to trace where almost any watershed begins and you will soon find yourself in philosophical quicksand. Is there a beginning, or is such a designation entirely arbitrary, an artifact of the human predilection for cutting things up into linear tracks, for taking the system apart so we don't even know that the whole thing fits together anymore? Precipitation comes down from the atmosphere in one place, percolates through rivers, creeks, streams, ponds, and so forth, and eventually empties into a big body of water—San Francisco Bay, in this case. But all the while that water is dispersing in other ways besides hurtling down cataracts. It is absorbed into vegetation, which respires, or breathes it out, at night. Fog like that enshrouding us now plays its rolling part in the passing off of water from one dimension of the system to another. Animals likewise are partly composed of water that comes and goes through us regularly as well. Water that gets to the bay or the ocean evaporates and is shaped into precipitation by wind and then it falls again at that spot you designated "at the top of the watershed."

Janet Klein, natural resources director of the MMWD and Williams' boss, puts the marker for the top of her watershed at the Eel River, all the way up in Mendocino County, hundreds of miles away. Eel River water empties into Lake Mendocino from which the Russian River issues. And that water makes its way to Tam via streams and creeks. All along the landmass for hundreds of miles, animals, plants, rocks, and soils collect, filter, and store water that eventually will be parsed out to nearly two hundred thousand residents of Marin. I am not one of them—as a San Franciscan I get my water mostly by way of the Hetch Hetchy Water System, the building of the central dam of which nearly broke the spirit of John Muir. Still, given its general provision of life through its tireless journeying, I agreed with Williams that we ought to thank the watershed and the least I could do was sleep on it.

I phoned home for emergency warmth supplies. My husband and son kindly delivered these to the Mountain Theater, where about twenty of us hardy souls set up for the night. I tried to get my husband to stay since he's very warm in a tent (my brave son stayed). "Ha ha ha, no way," he said, but he did give me his knit hat. There happened to be a lecture that night in the amphitheater. An astronomy professor from UC Berkeley across the bay talked about dark matter. The cosmic perspective turned out to powerfully trump the cold. I tried to stay awake as if repeatedly opening a door that insisted on closing. Those of us staying the night stumbled down the hill to our tents guided by bouncing headlamp beams. At about 5:30 the next morning, our evident unity in slumber with the sacred mountain was interrupted by the clanging delivery of porta-potties for a race that started at 8:00 AM.

At 2,571 feet, Mount Tam is the shortest of four mountains doing sentinel duty over the San Francisco Bay Area. (The West Peak of Tam once rose to 2,604 feet but in 1951 was shorn to

2,567 feet to make way for construction of a US Air Force station, since decommissioned.) There is Mount Saint Helena to the north in Sonoma (4,341 feet), Mount Hamilton on the peninsula (4,360 feet), and Mount Diablo in the East Bay (3,864 feet). These mountains all have gorgeous vistas and people who are devoted to them. But as Gary Snyder says, "Tamalpais is most exalted in California's remade human world."[5] He calls out the invitation there to contemplate the "mysterious beauty of oneness and separation." Like all watersheds, all mountains provoke appreciation of their distinct character while at the same time raising the question of where they actually begin and where they actually end. Mount Tam is not different from its precipitous peers in this regard.

Over the three years of the inventory, I became somewhat obsessed with what exactly Mount Tam embodies and exerts. Another citizen science project I participate in is the Hawkwatch across the bay in the Marin Headlands. This involves long hours of staring with binoculars facing the four cardinal directions. So in addition to contemplating Mount Tam from within its rocky folds, I consider it quite a bit from a distance. Most days with the least bit of clarity, both Mount Diablo and Mount Hamilton are also visible from the incredible perch of Hawk Hill near Sausalito. But Tam is a more immediate presence, on the north-northwest axis of the Hawkwatch, looking squat, dark, and magnetic. From this distance the mountain looks like the rain shadow it is. Storm systems coming in from the ocean bump up against Mount Tam, which causes them to release copious rain on its northern and western sides. This geological interference is thus responsible for its own weather system and for a significant portion of the water being sipped by residents.

Raptors sighted in the direction of Tam seem to disappear into it and arise out of it. Both on Mount Tam and on Hawk

Hill, the fevered human presence scattered around the bay seems incidental. Gary Snyder recalls a walk with Jack Kerouac, who immortalized Snyder as the character Japhy Ryder in *The Dharma Bums*, scampering like a mountain goat a little fast for the guy who immortalized the road trip. Referring to Snyder's lifelong preoccupation with ecological degradation and its discontents, he writes: "Stopped and thought about what I said to Jack about human history, remembered that Nature is inexhaustible, why should I fret at a few years of men?"[6]

Happy Trails

A gigantic cartographic representation of Mount Tam takes up the better part of a wall at Sky Oaks Ranger Station, and one day I asked Klein if this was the secret map of which I'd heard tell.

"No." (Looking at me like: "Don't even think about it.") "If you saw such a thing," she said, raising her eyebrows, "we would have to 're-educate' you or something."

The map I was looking at was dense with trails, but these were the official ones. There are roughly as many more unofficial trails made and used by hikers regularly. Klein and her colleagues at MMWD are responsible for monitoring them also, and there does exist a map of the secret side of the mountain as it were. I guess I'm not going to see it anytime soon.

The trails of Mount Tam are something of a living organism. The poet Robert Hass ponders the provenance of the first treks around Mount Tam by the Indians who lived here for thirteen thousand years before the white European rout, imagining their arrival from across the Bering Strait in pursuit of the megafauna they soon hunted to extinction. "The second thing that they must have done in the process is make trails." He sees the first human citizens on the landscape following animal paths that hew close to waterways. "That must have been the order: first the movement of water and then the movement of animals and then the movement of humans along the animal paths, and a geography would have emerged, the kind of physical and mental map that, for humans, makes a place a place."[7]

Of his initial walkabout on Mount Tam with fellow Beat poets Allen Ginsberg and Philip Whalen, Gary Snyder says, "As I confessed later, [we] basically made that ceremony, that ritual, that whole walk, up; based on some Japanese yamabushi background, a lot of mantra-singing in India, and a long experience of the crisscross of the Tamalpais trails with their many names, testament to volunteers and the admirable fanatics of years gone by." Like the MMWD itself, the trails of Tam have a citizen-focused provenance. In 1912 German-speaking union workers imported the hearty socialist, working-class, outdoor-loving ethic exemplified by Austria's Nature Friends and founded the Tourist Club. They hiked; they built trails; they had parties on the mountain. The New Deal brought more hands to the task, and the Civilian Conservation Corps essentially laid down the infrastructure for easy human congress with Tam, including the Mountain Theater itself, which we are still using today.

Before we headed out one Saturday morning, there was some talk about the circle transect. We reviewed the drill. Each team would go out with a botanist or someone who might as well be a botanist, they knew so much, and locate latitude and longitude coordinates provided to them by Williams. Around this central point a circle would be described using a tape measure and flags, and then the team would commence to hunting for species within it, also listed by Williams.

Clint Kellner explained to an inquirer that the circle was used to help track change over time. "If a fire comes through or the forest grows over it, the species composition will change in the circle, and we can keep track of that." Kellner is such as all long-term citizen science projects revere and treasure, an absolutely dedicated, practically-more-knowledgeable-than-the-leaders, regular volunteer. Kellner is a professional with expertise in botany and entomology (bugs), and he works for an environmental consulting firm. At the moment he was also describing how you track climate change impacts on vegetation. But Klein corrected Kellner. "We didn't set these up to track change over time," she said. "We're documenting what's in bloom, and not every single thing in the circle."

Here we land on one of the sticking points of citizen science. Science as it is pondered by academics is very much concerned with what question is being addressed in its name and how. There's a big difference between drawing a circle around a piece of earth and counting up all the species you find in it, and doing that again at a later date, and, alternatively, counting up just the species you find in a certain life stage—budding, blooming, or dying (more gently called senescence in botany talk). Williams and Klein designed our surveys to capture these stages of growth, and they set up species lists for us to find. In the three years of the survey, we collected more than 1,137 plants comprising more than 550 species. Adding photographic observations where no plant was collected, those numbers are 2,075 observations and more than 650 species. Logging more than 2,600 hours, 185 volunteers helped uncover 93 species not previously known to live on Mount Tam, including three grasses.

Nature is darn dense and layered and its composition changes over the seasons and over longer time periods. A useful survey has to be explicit about what snapshot of nature it is seeking to take, and then repeat that specific shot again at another time. Most of the three-year survey was conducted this way, but as we neared the end, we hadn't documented a handful of plants Williams knew were on the watershed and wanted to keep tabs on. "Some of these are rare, some bloom at weird times," said Williams. "The plot strategy has to be abandoned to find them. Instead we'll go drive around like on a safari to where they are and just get them. We'll lose the 'this place at this time' aspect of the rest of the survey and this will be more like 'at this place in the watershed you find this plant.'"

"Walking backward, and backward walking; walking forward and backward, has never stopped since the very moment before form arose," Snyder says, quoting Zen Master Dōgen. Generations of poets and artists have sung and painted the beauties of Mount Tam and sought to find a spiritual, symbolic context in the landscape. The form Snyder sees arising in the vegetation is the mandala, and he advises that one way to participate in this unity is to walk through the valley bottom and the mountaintop in the same trip. Maybe the circle transect has some unconscious reference to the mandala at least in the sense of a microcosm to ponder to better discern the whole. Snyder says that repeated circumambulations provide "a way to see the mountain with gratitude and attention in all seasons; in a steady circuit which is never the same twice. There is no exact repetition. 'Not even once,' someone said, 'can you step in the same river.'" Transects are an attempt to track precisely what happens as change over time unfolds. "You know what those circles are for?" someone piped up. "To keep the botanists in line!"

Dig This

"I'm not a very good citizen scientist," I said, realizing as we plunked down our bags at Rocky Ridge in May 2013 that I had forgotten a trowel.

"You're a *loyal* citizen scientist," said Terry Gosliner, directing me not all the way back to Sky Oaks but to his nearby car, where I found an extra trowel in the trunk. Gosliner is an expert in marine invertebrates but loves to botanize on the side—a professional in his day job and an amateur, in the root sense of the word, which means "to love," the rest of the time. Trudging toward the trowel I reflected grumpily on my loyalty and wrestled with an inner demon. It's simple. Citizen science is actually work. It's not always stimulating and fun. I didn't feel like photographing plants that day or writing down lots of numbers. I was bored already! It was already hot out! If I ate my lunch now I wouldn't have it later.

By the time I got back to the site with the trowel, another volunteer had taken the scribe job and it fell to me to collect the specimens. This meant digging up at least one representative of each species we were tracking, and if it wasn't rare but profuse, digging up lots of them. The collector had the job of arranging the specimens on a large newspaper-covered piece of cardboard. This took a bit of artfulness. Eventually the specimens would get an official mounting on special paper back at the academy. But we pressed them slightly to help transport them, and they needed to be squished in such a way that their basic forms remained discernable. The rest of the gang was busily finding, photographing, and writing. I started digging in earnest.

Digging, in dirt! Looking for another and another and another of the same seemingly anonymous little species of grass, this one with two leaves on its stalk, not three (that's another species!). Never before had I discerned that grass is not just grass, it's lots of grasses and they all have different identities and life histories! I loved all these grasses, wildflowers, and shrubs, as long as I could crawl around looking for them, dig them up, then arrange them artfully on newspaper.

And thus I found my botanical niche as a citizen scientist. Digging up plants and pressing them between newspaper and cardboard, I felt as though I were helping elucidate nature's tome. These representatives of the mountain were on their way to becoming herbarium specimens, and would thus be translated into the language of taxonomy by which the book of life is commonly read. How we read the leaves and read the landscape in general changes all the time, with new tools of perception, like crazy intense microscopes and big eyes-in-the-skies satellites; new ways of putting together earth history, like all this time and space and geography stuff; and also new insights into how things work together. But fundamental knowledge about both earth history and how things are working here today is still to be discerned in the presence and spatial distribution of species, including plants. With the green things held gently between my fingers, I was thinking, this is the sense of life that must have animated Alice Eastwood—in photographs of her from youth to old age one sees a joyous, deeply engaged intelligence. She beamed.

Eastwood came to the academy for the first time in 1891, a thirty-two-year-old schoolteacher with an avowed ambition to become a writer. But she also had an abiding passion for plants. According to Thomas F. Daniel, Eastwood was born in Canada and spent the early part of her life at the Toronto Asylum for the Insane. Daniel (himself a curator today at the academy) comments that this aspect of her past "in some way preadapted Eastwood for work in a major natural history museum."[8] Her nascent interest in greenery was fostered by an uncle with a big garden, and when a failed business venture of her father's necessitated placing her in a convent for a time, she was able to assist a priest tending his own assembled Eden there.

The academy would not perhaps be a major natural history museum were it not for Eastwood's clarity and heroism. Working diligently to build up the botanical collections, Eastwood made regular collecting trips around the West, self-financed with a small income from property she owned in Colorado, and lived simply in rented rooms on Nob Hill.

As we all see with our own eyes every day, individuals of the same species still come in a variety of shapes and sizes with different markings. A collector putting together a representation of a single species will designate a "type" specimen as the definitive example, while also collecting many variations on the theme, as we did in the MMWD inventory of Mount Tam. San Francisco buildings burned frequently in those days. Anticipating the likelihood of conflagration, Eastwood departed from the usual herbarium practice of storing specimens with the others of their species, and she made a special case in which she stored just the type specimens, reasoning these could be lowered out of a window in case of an emergency.

It was a prescient precaution. At around 5:00 AM on April 18, 1906, along with the rest of the Bay Area's residents, Eastwood was woken by the earthquake that marks one of the big before-and-after divisions in California history. Nob Hill, like San Francisco's other major protuberances, is bedrock, so little damage was immediately apparent at her home. This was not so down on Market and Fourth streets, where in 1891 the academy had built a new building in the heart of downtown with funds donated by the classically eccentric millionaire James Lick (who wore stinking rags his whole life).

Expressly forbidden to enter the unsafe museum, Eastwood did not hesitate. The building's marble staircase was destroyed, so Eastwood and Robert Porter, a young friend she had run into on the street, made their way to the sixth floor by way of the remnant iron railings. It is something to imagine, Eastwood climbing these in her long skirts. The type specimen case was damaged and couldn't be evacuated as planned. Eastwood piled specimens into an old work apron. Different versions of the story have her fashioning cords from ropes or curtains, but in any event she lowered them down to the first floor. In repeated relays she got 1,497 type specimens onto the ground. Daniel says she slipped her Zeiss lens into her pocket (the academy still possesses this item). Porter helped her secure a wagon and a driver, and the city started definitively to burn. "Behind her, in the bed of the wagon, was all that would remain of fifty-three years of botanical collecting and the largest herbarium in the western United States," writes Daniel. The fire leveled the city, consuming not only everything left in the academy but also Eastwood's Nob Hill home and her possessions therein. She moved the specimens twice before they were quite clear of the encroaching fire and ultimately stored them in the vault of the Crocker Bank.

Of the experience Eastwood wrote: "My own destroyed work I do not lament, for it was a joy to me while I did it, and I can still have the same joy in starting it again. . . . The kindness of my friends has been great. . . . I feel how very fortunate I am; not at all like an unfortunate who has lost all her personal possessions and home."[9] Eastwood's outlook is that of the constant beginner, ready to be beguiled anew every day by something as evidently simple as a plant—no wonder the Buddhist-Beat poets took a shine to her.

Daniel calls the 1,497 rescued type specimens "a physical and a symbolic icon around which the department (and the institution) could rally for years to come." As in a fire-adapted plant, Eastwood's growth both personally and on behalf of the academy was stimulated by the conflagration. The academy curators were advised to find work elsewhere, and for the next ten years Eastwood set up botanical shop at UC Berkeley.

The 1906 earthquake seems like faraway history to us, but in the long geological time frame, not so much. The San Andreas Fault is one of the very central narrative lines upon which San Francisco history, botanical and otherwise, has been writ. Drawn on a map, the San Andreas Fault looks like a seam comfortably traveling within an edge of land from the mountains of Southern California extending north. The fault gets closer and closer to the ocean at Santa Cruz, and then it looks like whoever was operating the sewing machine got distracted and started texting while stitching. When it gets to San Francisco Bay it veers between land and water, crossing west of the Golden Gate between the mainland and the Farallon Islands. Back to land through Marin by way of the Bolinas Lagoon, back out at Tomales Bay, back in to land at Sonoma County, out again, then in again at Mendocino County, from which it departs into the Pacific north of Point Arena. This description, of course, is entirely anthropocentric. In the longest and deepest processes of earth, water and land are in a constant state of flow relative to one another, and a coastline is a temporary boundary.

The separations made by the fault line have as much to do with age as space and have a definitive impact on the plant life of Mount Tam. Even geologist Doris Sloan said the geological complexity here "defies order and reason."[10] But we can take a stab at some of the main factors. First there's the whole enchilada of California's long ring of mountains, made by "faulting, folding, uplifting, and volcanism,"[11] and then the definitive subduction of the Pacific Plate under the westward-moving North American Plate. The eastern side is made up mostly of the Cascade Range and Sierra Nevada; the Klamath Mountains and the Coast Ranges make up the western side; and the Transverse Ranges and Peninsular Ranges extend down to Baja. The Central Valley is made up of rich soils resulting from the glacial runoffs of many an ice age.

At Mount Tam the pas de deux between plates has thrown some old rocks up among some newer ones. East of the fault the Franciscan Complex, hailing to the Jurassic, is mostly sedimentary rocks, which means they were formed at the earth's surface from skeletons of plants and animals and out of fragments from other rocks. Perhaps these were the first California immigrants, because the Franciscan Complex rafted here via moving plates. Riding my bike up the very very very steep grade to Hawk Hill in the Marin Headlands, I have frequently been mesmerized by walls of what Sloan calls "folded Franciscan radiolarian chert." Radiolaria are protozoa that leave behind sci-fi-looking mineral skeletons—which you don't discern with the naked eye, but you do get to ogle a linear pattern in the rock that looks like rough-edged pottery.

What grows where has to do with what it's growing in, what kind of substrate, and the attendant climate. In a kind of general, holistic way, Gary Snyder puts his arms around the geography and plant life of Mount Tam: "The natural ecosystems—plant zones—life zones—mini floristic provinces—are a great and com-plicated mosaic, from the ocean-facing foggy and windy side to the different faces on the dryer sides, each with its own logging, ranching, and fire history."[12] In 1889 C. Hart Merriam proposed a system for organizing where we find plant and animal com-munities according to elevation and attendant climate differ-ences. The life zone concept is challenged in Marin, because the climate, substratum, and topography are so varied and in mishmash shape. Very different soil and moisture conditions exist side by side and that results in very different plant group neighbors.

So plants that are close together spatially are separated quite widely by time—the time in which the rock below came to the sur-face and eroded enough to support vegetation. The redwoods on Tam all hail back to the Franciscan Complex, and thus an outline of where they appear is like a hologram of ancient Marin, which has since been altered and revised by newer rocks supporting newer plant life. The distinctive Sargent cypress is also limited to Franciscan rocks. With sedimentary rocks, the other major sorts of hard things are igneous, formed as molten rock cools, either underground or once it erupts to the surface by way of a volcano. Then there's metamorphic rock, which is rock of any origin cooked by heat, squashed by pressure, or both.

Poking up from the sedimentary rock on Mount Tam are igneous rocks that have metamorphosed into serpentine, California's state rock. That Hopi elder of course knew what he was about in directing Gary Snyder and his beatnik crew to place a beginning marker for their journeying at a serpentine outcropping. Serpentine is ancient and harsh stuff, with few nutrients, which has ironically been the saving grace of the hardy plants that do grow on it. Though some might consider them homely weeds, others value these plants as jewels from the mists of time, such that make Tutankhamen's treasure look like it was made yesterday on a geological assembly line. Plants adapted to serpentine are like living history books in and of themselves. Their successive generations trace ancestry at least to the Jurassic and probably earlier.

Fire and Water

As the business school adage goes, if you can't measure it you can't manage it. And managing vegetation on land sustaining a municipal water supply is important indeed. "Our nineties plan for the watershed was all about fire. The ideas about fire have changed quite a bit," Williams told me. "Previously we thought that over big areas, ridgetops act as big fuel breaks," which theoretically bring a raging inferno to a halt. "Now in California we know our odds of stopping a fire are about equal to our odds of stopping an earthquake. There's so much fuel that once it gets going it's going to be big. The right winds in a fire pick up enough steam that the fire creates its own weather."

To add another layer on how we are viewing the landscape now, consider that fire has also had a regular historical role in editing and re-editing this local chapter in "Nature's infinite book," to quote the soothsayer in Shakespeare's *Antony and Cleopatra.* What she would discern in the palm of Mount Tam would be a cycle of fire, in the past mostly moderated by Indians who, as we know, burned on purpose, which left the redwoods mostly unfazed and even jazzed by the hot pruning action, and took out the Douglas fir. Doug fir is killed by fire as redwoods are not. It is then succeeded by bushy cover that makes it possible for new Doug fir seeds brought by wind, birds, small mammals, herbivores, and so forth, to germinate. Fire acts as a propagator of species and also helps disperse them. For thousands of years, the soil and plant life here was adapted to the California Indian burning cycles, and for several hundred years, all burning has been suppressed. One result is a buildup of duff and debris shed by the plants, a real and present hazard to the community in the event that on one of our hot, dry days a spark will fly.

The small burgs of Marin are arrayed around and on Mount Tam, and from the different MMWD survey sites we volunteers variously looked down on the hippie-ish towns of Fairfax and San Anselmo, where I had attended Wiccan campfires at the summer solstice, and Kentwood and Ross, where I had been to dinner served by a butler and staff at a mansion built on "the last really fine estate in Marin." Thinking of the fellow who owns that estate, who of course has to tell you his is the best, and remembering the artwork on his walls, I thought he wouldn't be so keen on a big fire hazard.

I said to Williams, "Your locals don't like that about the fire, do they?"

"No."

Public hearings about the water district often devolve into shouting matches about getting rid of the French brome invading the watershed. French brome is an invasive species, not only disrupting the native plant life but also serving as a fire accelerator. Williams and Klein would very much like to get rid of the French brome. Teams of students from all over Marin, and other volunteers, come out regularly to pull it from the ground. Trouble is, they can't possibly keep up with it. As Klein told me, "You have to poison it."

"A problem," I replied with the confidence of hitting on the very obvious, "when you are dealing with a water supply."

"There's a dosage at which point the herbicide becomes poison," she said, "but we wouldn't use it at that level on the watershed."

Klein is in an impossible spot in the matter of French brome. I asked her about the possibility of just living with it. "What about the argument that we are all aliens here, we are all invasive species, and it's not fair to label a plant as noxious just because it comes from somewhere else?"

"I don't care where you came from or whether you're native or not, I care how much money you're costing me," said Klein. "French brome propagates so fast it costs too much to keep up with it. It's getting all over the mountain and the whole thing will go up like tinder the minute we have a fire."

And that's not an *if*, that's a *when*.

The Global Garden

Before arriving in California, Alice Eastwood spent ten years teaching at East High School in Denver, Colorado. While her legacy is most expressly connected to California flora, it was to the Rocky Mountain wildflowers that her passion first tended, and she eventually self-published *A Popular Flora of Denver*,

Colorado. Seven years after Asa Gray climbed the mountain that thereafter has been called Grays Peak, Eastwood followed suit, and did so every summer after school let out, collecting specimens. In July 1887 none other than Alfred Russel Wallace himself, already famous for his contribution to the theory of evolution by way of natural selection, came to botanize on Grays Peak. He was continuing his global investigations into the distribution of plants, how and when they got to be where they are, and wanted to compare Western and European vegetation. This was in support of his concept that the continents of the Northern Hemisphere had once been a continuous landmass. Wallace consulted with the principal of East High School who told him only Alice Eastwood was qualified to guide him, and so she did.

She was five feet tall, she was pretty and lively, and she wore long skirts as women of the day had to, but Eastwood spent years botanizing all over the West with men to whom she was not married or otherwise related. (She narrowly escaped marriage twice.) In the matter of Wallace, the two took a train to Graymont (Asa left his moniker all over the place), spent the night in a two-story log "hotel," and then hiked to Silver Plume Mine the next day. They were accompanied from there by the manager of the mine, and the three of them spent the next night in a hut, with a blanket dividing the one room into male and female quarters. Making the summit of the peak the next day, Eastwood and Wallace were reportedly so stunned by the effusion of flowers that they lost their way up there. Making a continental hypothesis, Wallace compared the "alpine gardens" with similar flowers in the European Alps. (He collected plants to mail to Gertrude Jekyll, the horticulturist responsible for exploding the notion of what a garden should look like—the teeming, British garden full of drowsing rose heads was her conception and still rules much garden design today.) In the ten-year hiatus between the burning of the old academy and establishment of a new institution in Golden Gate Park, Eastwood traveled widely and included in her sojourn a trip to Wallace's home at Orchard House, where Wallace and his wife welcomed her like family.

Making her mark in San Francisco, Eastwood expended equal energy on garden clubs, and what we call outreach today, as she did on building the museum collections. She just loved flowers and thought everybody must. She established an ongoing educational flower show in the entrance lobby of the old California Academy of Sciences building in Golden Gate Park, the longest continuous floral exhibit in the world for many years.

Who knows why he thought it was useful or relevant to do so, but Joseph Grinnell, founding director of UC Berkeley's Museum of Vertebrate Zoology, let Eastwood know of a certain "distinguished professor's criticism of the mismatched containers" Eastwood was using in her displays. She responded thusly: "If I could do it I'd have a different kind of receptacle in color and form for each kind of flower as I abhor uniformity just as nature does. So long as I run that flower show I do it according to my ideas. . . . You know my opinion of the eastern professor, some hide-bound stickler for system where system is not the desirable feature. You need not write to me again on this subject."[13] Wow, girl.

In the late 1930s, "old Mrs. Wilkins," a janitor who helped her arrange the flowers, took note of a young German immigrant's frequent visits to the displays and his careful documenting of observations. Mrs. Wilkins sent Eric Walther to talk to Eastwood, who encouraged his scholarship and gave him keys to the herbarium so he could consult the academy's library. Eastwood kept Walther in mind as she brooded on one of her dreams, to establish a botanical garden in Golden Gate Park. The park's first supervisor, John McLaren, wholeheartedly agreed with her on the need for a garden within the park that would show the world the amazing flora of California, at the time pretty much new to science. They were also keen to display the abundance of plant life from far-flung places that could grow here, thanks to the climate. As early as 1889, McLaren identified a location for a botanical garden in the park. He supported an 1898 bond issue to establish an arboretum and botanical garden, but it failed to get a two-thirds majority. The site was planted with a grove of coast redwoods around 1900, and these trees have grown into giants that today grow along what has become the Redwood Nature Trail.

The garden finally became real in the midst of the Great Depression, and workers employed by the Works Progress Administration helped with construction. Walther was working down the peninsula on an estate, and when Eastwood learned he had applied for a job in Golden Gate Park, she had a word with John McLaren. Walther got the job as first director of the garden. His academic perspective, schooled by Eastwood, influenced his progressive plan for organizing the garden's plants. Many contemporary botanical gardens are based on big beds of the showiest flowers, but San Francisco's is based on historical geography, the very thing Darwin and Wallace pondered while hatching their ideas about evolution, and the very thing Wallace was investigating in Colorado guided by Eastwood. The San Francisco Botanical Garden synopsizes the earth history processes that brought us vegetation and different kinds of it.

Today it features an Ancient Plant Garden, which helps explicate how plant life itself evolved and the critical role it plays in how the rest of life, including mammals like us, are able to live on the planet. Three billion years ago cyanobacteria got the oxygen-producing show on the road. Algae evolved and then lichen, moss, and liverworts came along–there are no interpretive signs for these at the garden, but the careful explorer can find plenty. The Eocene brought us a world dominated by flowering plants, including the garden's iconic magnolias. Magnolias evolved before bees did, and some are still pollinated by beetles. The garden's California coast redwood, *Sequoia sempervirens*, and the dawn redwood, *Metasequoia glyptostroboides*, bear witness to an earth as it existed before we were here. While the redwoods and the Garden of California Native Plants most express the origins of plant life in the Laurasian, or Northern, Hemisphere, part of an original megacontinent called Pangaea, the garden is largely devoted to significant collections of Chilean, New Zealand, Australian, and South African gardens–these places were all located in the portion of Pangaea that extended into the Southern Hemisphere, called Gondwanaland.

The Eye of Heaven

Eastwood couldn't have anticipated that one day the garden would be a useful template for helping people understand the impacts of climate change and habitat disruption. The garden is located on the Pacific Flyway. As something is always in bloom here, it is a favorite fuel stop for those migrating north-south and south-north, as well as for many full-time resident birds. A number of factors, including different temperature and precipitation ranges and the fruiting and flowering timetable of various plants, influence which species use the garden and when. The close intertwining of bees, other insects, and pollinating birds with host plants is the daily business of the garden. Since it has a big collection of native plants, the native bees that come to these are likely to be the replacement troops as honeybees from elsewhere are declining in abundance. Small, cultivated places like botanical gardens have already become sources for rewilding the rest of the globe.

Some wilderness advocates blanch at the suggestion that we consider ourselves gardeners of the world. The idea smacks of hubris. Philosopher Eileen Crist says the problem begins with the practically unconscious consensus that puts *Homo sapiens* in a special place above other life-forms. Crist references the term *oecumene*, which the Greeks used to describe the known world—the world known to humans, exclusively. Thus by language alone we have inherited an idea that the "real" world is the human world. Crist says this results in an "existential apartheid" that justifies human interests above and at the expense of most other life-forms.

Crist's description of the self-interest embedded in Greek thought is reflected by Mike Ghiselin in "The Individual in the Darwinian Revolution."[14] Aristotle gets billing as an early citizen scientist, but his case perhaps makes a better argument for leaving science to professionals. Ghiselin points out that in Aristotle's works on natural history, "we find a wealth of impressively accurate observations.... Yet mixed in with what are now considered facts, we keep finding curious modes of explanation intruding into the data." Aristotle opined that round eggs hatch into males and elongated ones into females; the more perfect sphere produces the more perfect sex. Aristotle said "up is more noble than posterior," and right is better than left. Ghiselin makes a quick cut and thrust: "Aristotle walked erect, faced toward the front, defecated to the rear, and, we may suspect, was right-handed. Indeed his entire cosmology seems remarkably appropriate to the prejudices of a Greek aristocrat." In Aristotle's cosmology, we find God in self-contemplation presiding from the top over a rational world; man looks back up at the eternal and the divine, and animals and plants "exist for the sake of man."

And that's not all. This way of seeing nature corresponds to Aristotle's "doctrine of priority" in which, for example, a master is "prior" to his slave, who thus exists for the master. Among other things, Aristotle lays down philosophical justification for colonial expansion, the energetic activity by which centuries of Europeans dominated and often destroyed other cultures. Crist says: "'Gardened planet' is a euphemism for colonized Earth. And humanity is not penning another interesting chapter of natural history, but heralding the end of a sublime one."

But today the reality is that climate change touches everything; there's no escaping it. Places like botanical gardens, which can support species that are rare and endangered or even extinct in the wild, have become repositories of genetic material we are going to need to regerminate diversity on the planet. We are going to have to re-create the Garden. We don't have any choice but to consider ourselves gardeners charged with caring for creation. Author and Stanford University professor Robert Pogue Harrison says gardens "visibly gather around themselves the spiritual, mental, and physical energies that their surroundings would otherwise dissipate, disperse, and dissolve." He points out that the very definition of a garden is essentially a kind of boundary, and gardens are "gateways . . . through which you may be called upon or visited, without moving from where you stand." Citing among other instances Dante's portrayal of Beatrice's epiphany in the Garden of Eden, he comments that "gardens in the human imagination often figure as the sites of visions."[15] Or breaking through, as Ed Ricketts would have it.

Fundamentally, climate change busts through all these boundaries and interpenetrates the special place of the garden and what surrounds it. This reordering of things is profound on every level and even changes how we read literature. A trip to Shakespeare Garden in Golden Gate Park (across the street from the botanical garden) showcases again the mentality and care of Alice East-wood, who dreamed it up and saw its construction through. It was just in Eastwood's line to fastidiously pore over the plays and the sonnets and to give physical reality to poetry she loved. Consider what climate change has done to Shakespeare. "Shall I compare thee to a summer's day?" begins one of the greatest love poems in English literature. But what is a "summer's day" now? "Sometime too hot the eye of heaven shines," says Shakespeare, but now it is pretty much always shining too hot. "The eye of heaven" implies an implacable divine observer sussing up the doings down below. It is a wonder we don't feel shame about climate change—we should. The sonnet says as long as its verse exists, so the beloved remains alive, to be invoked by "nature speaking," as Campbell put it. But the metaphors holding this dearest one in linguistic embrace, from "the darling buds of May" to the "summer's day" itself, are out of whack with actual nature. The darling buds are probably coming out in April. The summer's day is in many places all year round, where it used to be confined to one season out of four. So now must be added to the revelations in the garden, as Hamlet put it, "What a piece of work is a man."

The Great Unzipping

One day out on the watershed I was charged with digging up a bouquet of *Plantago erecta* (California plantain), and as I crawled around sifting through stalks barely discernable from one another to find this hearty native, I marveled that the tiny stem and truly unassuming flower is a rock star in the history of evolutionary biology. *Plantago erecta* grows in a lot of different substrates but notably in serpentine. When the expeditions to collect Eden went out from Europe hundreds of years ago, those early botanists intuited that deep truth would be readable in the leaves. *Plantago erecta* has just such a story to tell.

While the protagonists in this case are two famous scientists, Paul Ehrlich and Peter Raven, the work they did to quantify the concept of coevolution depended entirely on the work of generations who came before them—mostly amateurs, or citizen scientists, those obsessed with lepidoptera. Paul Ehrlich is perhaps best known as the author, with his wife Anne, of 1968's *The Population Bomb*, but his status as one of the foremost scientists of our time originates from research not on human but on butterfly populations. Ehrlich arrived at Stanford in 1959 and soon set up research transects at Jasper Ridge, the university's nearby research station and 1,189-acre reserve, initiating a mark-release-capture study of areas where *Euphydryas editha bayensis* butterflies were known to tarry. His ambition was big: He wanted to establish the colloquially known Bay checkerspot butterfly as a model for studying ecology in the way *Drosophila*, or fruit flies, are a model for studying genetics. Ehrlich and his Bay checkerspots have contributed foundational information on population dynamics and genetics. He is also the paterfamilias of several subsequent generations of butterfly researchers trained by him at Stanford. Over decades they have indeed established the Bay checkerspot as a model system.

Ehrlich's first big splash in academia was a 1964 paper he coauthored with botanist Peter Raven, "Butterflies and Plants: A Study in Coevolution." "Peter and I worked in the same building," Ehrlich told me one day over lunch, gesturing toward what is now the university's art museum. "We used to have coffee or lunch every day, and we have the same sense of humor. One day I said, 'Peter, I have *Euphydryas editha* butterflies that feed primarily on *Plantago erecta* but they also feed on owl's clover. It's absolutely nuts." *Plantago erecta* is a rather simple-looking herb while owl's clover is a pretty wildflower. "Peter said, 'No, it makes perfect sense.'"

I agreed to meet Ehrlich at 11:15 AM so we could get served at his favorite campus café ahead of the hordes. As we lunched a student approached stammering and blushing and requested a photograph. "Oh my God, my parents are going to be thrilled," he said, as a buddy snapped away with a smartphone. I myself had a case of *intimiditus* on my way to meet Ehrlich for the first time, which is kind of funny now. Paul Ehrlich is a hail fellow well met, conversant on just about any subject, and a fantastic lunch companion. While pleased I was writing about coevolution, he made a request. "Would you please let your readers know that the burgeoning human population is the biggest threat to biodiversity today, and indeed to all of life?" Check.

Peter Raven, not the same kind of household name but equally revered in the world of science (perhaps more so in some circles), is similarly charming, cosmopolitan, and deeply concerned with saving nature. Raven hails from San Francisco, and his ardor for greenery was doubly inflamed when at age nine he presented Alice Eastwood herself with a specimen. "When I first met her in 1946, she was near her ninetieth birthday," Raven recalled. By then she was as august a presence as the California Academy of Sciences has ever known. The young Raven showed her a collection of a small wood rose and she responded, "How happy I am that you collected it with fruit. Now it is really quite simple to identify it—see, it has no hips."[16]

Constant Gardeners
Ehrlich and Raven put one in mind of the earnest scholars who sent out expeditions to find the source of creation, only now they are trying to find ways to save it. Both are highly involved with efforts to stem extinction. Ehrlich heads up the Millennium Alliance for Humanity and the Biosphere at Stanford University, which basically wants to save the world. With two other

scientists, Raven is a principal author of the pope's encyclical on climate change. Covering the Vatican's unveiling of this unprecedented document in July 2015, Naomi Klein reported that in prescribing what we should do about degradation of the earth, the encyclical references "care" dozens of times and "stewardship" only twice. "This is no accident, we are told," Klein writes, referring to the keynote address at a gathering organized by the Vatican. "While stewardship speaks to a relationship based on duty, 'when one cares for something it is something one does with passion and love.'"[17] This of course is the position of the amateur.

Once upon a time these two silverbacks were young primates figuring out evolution. Raven told Ehrlich that in fact *Plantago erecta* and owl's clover are closely related, despite the fact that they look so different, and that of course the butterfly larvae would respond similarly to both plants. "We had this eureka moment," Raven told me. "We both realized that yes, plants are linked in groups by chemicals in them, and that meant something." Until then, a widespread assumption was that chemicals in plants that don't contribute to the plant's metabolism or development must be excretory products.

"It dawned on us that this was crazy," Ehrlich said. "Why would any organism make an expensive excretory product and then not excrete it? Plants can't run away, so they have to defend themselves against predation some other way. We had the simultaneous hunch that the chemical properties of plants may have evolved to make the plant taste bad to larval butterflies."

The process, called "coevolution," by which two closely linked species develop traits in tandem with one another, was intuited by Charles Darwin. Confronted with the Madagascar star orchid in 1862, Darwin mused that there must exist an insect with a proboscis long enough to penetrate the plant's eight- to fourteen-inch spur to obtain its nectar. The idea of such a bug was roundly ridiculed. In 1862, George Campbell, eighth Duke of Argyll, argued in *The Reign of Law* that the complex plant must have supernatural origins. Alfred Russel Wallace once again buttressed the thinking of his better-known peer. In "Creation by Law," Wallace articulated a step-by-step process by which the survival success of the Madagascar star orchid would be matched by the success of an insect with a commensurate proboscis. In 1903, what has become known as Darwin's hawk moth was discovered in Madagascar. The moth unfurls its long proboscis to go for the nectar and rolls it back up again when not in use. Darwin was dead by this time, but Wallace was still around and got to crow.

In the case of *Euphydryas editha*, the question of simultaneously developing traits was about not pollination but herbivory. Butterflies lay eggs on particular plants that are their favorite foods; the eggs develop into caterpillars and begin munching on their host. Some plants develop toxins to deter being eaten.

"Butterflies feed on lots of different types of plants," Raven explained, "but most butterflies feed on only a narrow group of plants with similar chemical protections. That was very interesting. We visualized a coevolutionary arms race in which a group of plants would develop a certain chemical defense, protecting them from all insects for a while. Over time, some butterfly species evolved the ability to break down the poisonous chemicals, or store them in their bodies. They then became very specialized on the plants with that chemical defense, because there was little competition for them. So back and forth, plants develop new chemical defenses and then insects develop a way to deal with it in a stepwise evolutionary progression."

Most ecological research occurs at least partly out of doors, but in this case Raven and Ehrlich dug into their idea exclusively inside, in Bay Area libraries. The lion's share of butterfly food preferences had been documented by amateur collectors in journals. Likewise, the chemical components of plants were well-documented. Raven and Ehrlich laid out a giant list of plant food choices for each butterfly family they studied, all together between 46 and 60 percent of all butterfly genera—such a big number, they wrote, that it is "highly unlikely that future discoveries will necessitate extensive revisions of the conclusions drawn from this paper."[18] Then they compared butterfly group with plant group. Where the plant groups had diverged, so had the butterfly groups. "The evolution of secondary plant substances and the stepwise evolutionary responses to these... have clearly been the dominant factors in the evolution of butterflies."[19] In other words, when the plant changed its composition, the butterfly changed to accommodate it.

Butterfly Whisperer

The Bay checkerspot enjoys marquee billing as the species that suggested a major ecological concept, coevolution. Like a movie star who crashes and burns, the butterfly has also become a poster child for human-induced extinction. Continuing to study the Bay checkerspot at Jasper Ridge, Ehrlich and colleagues were helpless to protect their own study subject from disappearing forever from this place. When drought reduced the number of Bay checkerspots at Jasper Ridge, under historical circumstances it is likely their local population would have been renewed by immigrant butterflies from nearby. But the closest populations of Bay checkerspots were eliminated when adjacent habitat in Woodside was bulldozed for development. Thus there are no more Bay checkerspots at Jasper Ridge.

Ehrlich and others purposed this sad happening into knowledge about what metapopulations have to do with extinction.[20]

We frequently use the term *meta* in a colloquial way to indicate a big collection of similar parts, or to indicate more than the sum of parts, something "beyond," as in *metaphysics*. A metapopulation in biology touches down on both definitions. A metapopulation of butterflies is an aggregate of separate populations that have some dependence on each other. At first glance one might think it sad but not the end of the world when a single checkerspot butterfly population goes extinct; at least there are more populations out there. However, the situation is worse than that. A metapopulation needs a certain number of smaller populations to keep the aggregate going—lose too many and everybody goes poof.

Stuart Weiss is a scientist who remembers trying valiantly to save the Jasper Ridge checkerspots with Ehrlich. For more than thirty years now, Weiss has dedicated himself to saving those that are still around, notably in San Jose's Coyote Valley and in San Mateo's Edgewood Park. "Sounds like you've been led around by a pretty flitterer," I said to him. "Yep," he confirmed. "I'm totally manipulated by that butterfly. And honored to be. We are both part of a web of coevolutionary relationships. The butterfly is totally dependent upon us not doing the wrong thing."

Just south of San Francisco, Weiss and colleagues from the Creekside Center for Earth Observation are working with citizen scientists to boost the Bay checkerspot butterfly population at the Edgewood Park and Nature Preserve. They are also working to increase the numbers of the San Mateo thornmint, also an endangered species. On a spring day in 2014, I watched Weiss' volunteers as they counted thornmint stalks in a transect. They were so focused on the elusive little plants, they looked like surgeons bending over a green operating table. It's hard to tell whether rescue efforts are working or not, because the thornmint population fluctuates from about fifty thousand individuals in a good year here to less than five thousand. It grows only on serpentine soil, making it an endemic plant.

Traffic streamed by on its ceaseless course. Looking down over Interstate 280, which I myself had traveled on to meet him, Weiss called my attention to the differences in the vegetation on either side of the road. Where we stood, the hillside was getting full sun, and Weiss told me the plants here were similar to those you find in Southern California—chaparral, sage, chemise, and deerweed. "It's a different microclimate over there," Weiss said, pointing to the other side of 280. Looking now with focus and attention, I could readily see the landscape was a denser tangle of darker green. "The mountains are full of oak woodlands and Douglas fir," he told me. "But look." He pointed out a stand of chapparal. "That's ready to take off," he told me. This one splotch of brownish bush would expand given drier, hotter conditions.

Weiss and his colleagues quantify something called "climate space," and to get my mind around the idea of that, it helped me to think of paintings by Seurat or Van Gogh, where every little space has its own color. In the case of the actual landscape, every square inch has an aspect to the sun and a relative elevation. Vegetation is directly responsive to these conditions, which might be called mini-microclimates. Lots of further inputs, like the kind of soil and rock you find in a spot, also influence the climate space and what grows in it.

One reason the difference between the sides of 280 are so stark here is that the road runs along the San Andreas fault line. Those two converging tectonic plates, the Farallon and the North American, mashed up about 35 and 165 million years ago, and the resulting serpentine soil has ironically provided a haven for some endemic plants. Serpentine soil is nutrient poor, and not too many things can grow in it. Among the plants that do grow in it are *Plantago erecta*, which is the host plant for Bay checker-spot butterflies. (I asked Weiss why there aren't any Bay check-erspots at Mount Tam even though there's plenty of serpentine soil and *Plantago erecta* there. He told me there may once have been, but it's probably not possible to ever really know.) Scores of citizen scientists have helped Weiss transfer Bay checkerspot larvae to *Plantago erecta* stalks here.

Weiss has drilled deeper into the coevolution between but-terfly and plant to quantify and document the effect of sunlight on the relationship. In the early part of his career, with Ehrlich as his advisor, "I worked on the butterfly phenology at Jasper Ridge," Weiss recounted. *Phenology* is from the Greek "to show" and describes nature's scheduled appearances such as flowering, fruiting, dying, and migrating, and also nature's hiding, hiber-nation. In other words, phenology is about the timing of natural events. The Bay checkerspot keeps a schedule in close parallel to that of *Plantago erecta*. In November and December, the but-terfly larvae come out of diapause (dormancy, huddled under roots and in cracks in the ground) and start to eat. The furry black caterpillars lie around like true Californians in the sun, soaking up rays to raise their body temperature, and in Febru-ary through March go through their chrysalis phase in the grass. Emergence and the flight season occur through April; once they awaken to their full form they get busy and mate. The timing of the solar rays on sunny or relatively cooler slopes impacts the life cycle of the *Plantago* and thus the butterfly.

Perversely, the Bay checkerspot has another problem here, which is that nitrogen in the cars' exhaust acts as a fertilizer. Overloading natural systems with nitrogen not only from vehi-cle exhaust but from agricultural runoff is a devilish problem that thwacks growth patterns. In some places, nitrogen poison-ing causes ecosystem functions to totally collapse. In this case, invasive Italian ryegrass feeds on the nitrogen, growing up so high it eventually suffocates the *Plantago erecta* beneath it. Butterflies trying to get to their favorite food go hungry and die. "But we figured out that right here, anyway, we can help the butterflies by grazing or mowing the grass," Weiss told me. Every couple of years the county comes out and cuts down the grass, giving the *Plantago erecta* some breathing room. On bigger swaths of land, it is more cost-effective to put cows on the landscape and let them eat the grasses periodically. And the butterflies will rise up like dark angels—for the moment, anyhow—their resurrection completed.

Wings of Desire

Raven and Ehrlich set up a framework for understanding that the evolution of one species is frequently intimately tied with the evolution of another; at the time, they could have no idea that in a few short decades, such time-hewn relationships among species would start to come unzipped thanks to climate change.

"A really important dimension of the Ehrlich-Raven coevolution paper is that it puts the ecological time scale within an evolutionary framework," David Inouye told me. A slight man with dark hair and tinted glasses, Inouye could be an angel in a Wim Wenders film, looking carefully over the shoulders of butterflies and bees rather than anguished *Homo sapiens*. But in fact he is a biology professor at the University of Maryland, and for the past forty-five summers he has conducted a field study in flowering phenology at the Rocky Mountain Biological Laboratory in Crested Butte, Colorado, a favorite spot for long-term ecological experiments among scientists from all over the world, including Raven and Ehrlich.

"There's always an expert around," Inouye told me. "I can ask about a particular butterfly and there's always somebody there who knows everything about it." Crested Butte is "the wild-flower capital of Colorado," and although initially focused on bumblebees and hummingbirds, Inouye soon "got interested in the resources," or the flowers on which the insects and birds depend. "We look at what species is growing in different habitats, and document the date of first, peak, and last bloom," a.k.a. phenology.

Inouye's work and its far-reaching conclusions have been made possible by what the professionals call a "shoe box data set," or the collected work of one exemplary citizen scientist, Billy Barr. Inouye described Barr as Rocky Mountain Biological Lab's one-time "dishwasher, librarian, business manager," and current accountant. While Inouye, along with forty or so scientists plus forty or so more grad students, spend hunks of summer at RMBL (fondly known as "Rumble"), Billy Barr spends all year in the vicinity. Partly because he's pretty much alone in the winter months (and calls it "boring"), he has been keeping close tabs on nature. Barr has kept a weather journal every single day for more than forty years, noting quality of light, cloud cover, and wind. He has established the world's most comprehensive database documenting naturally occurring avalanches. And most importantly for Inouye and other researchers parsing out how biotic (plant and animal) and abiotic (climate) interactions are affected by climate change, Billy Barr has for the past four decades measured the depth and weight of snow.

When Inouye's study began, he had no idea that with time his work would become among the most important to document the effects of climate change on ecological processes. "It turns out that the primary determinant of phenology in the Rockies has to do with when snow melts," Inouye told me. "When there's snow on the ground the insects and hibernating animals don't come out; when it melts, they do." Inouye and his colleagues have correlated progressively earlier snowmelt dates with increasingly warm temperatures associated with climate change. "The snow melts and the flowers are triggered into blooming," Inouye said. "But the pollinators haven't arrived yet; the migratory hummingbirds are on a different schedule. And although the blooming dates are coming earlier, the date of the last hard frost hasn't changed, so between June tenth and fourteenth, the hard frost comes and kills the buds and flowers. When the pollinators do arrive, there's nothing for them to eat." Among other things, Inouye said, "soon Crested Butte is going to have to change the date of its wildflower festival."

Other species affected by this thwacking of nature's historical coordination include the marmot, a cute hibernating mammal that is kind of fat and getting fatter. "During the past forty years," Inouye reported, "marmot hibernation time has decreased by forty days—one fewer day per year." With longer time spent out and about, the marmots eat more. The marmots emerge when it's warm enough to do so, and recently there's still been snow on the ground. The result is a fur ball highly conspicuous against the white snow, which makes them easier to pick off by coyotes. The snowshoe hare has been experiencing a similar disadvantage. This rabbit's fur changes from brown to white to capitalize on camouflage in the snow, but the color shifting is timed to temperature changes. Now those temperatures are changing ahead or behind their historical schedules, and the hares' coloration is becoming mismatched with the snow. This can mean brown hares against white snow or white hares against brown and green ground. Either way the hares become easier pickings for Canada lynx.

And to add a devilish feedback mechanism into the mix, massive dust storms originating mostly in Arizona are speeding the snowmelt in Colorado. The southwestern desert floor is protected by a cryptobiotic crust, which is made up of lichens, algae, bacteria, and fungi and is a lot more delicate than the name sounds. Increased all-terrain vehicle use, as well as the usual disruptions of development, are breaking up the crust. Huge dust storms result and settle on the snowy mountaintops up north; the darker color of the dust changes what's called the snow's albedo, or the quantity of light it reflects. It gets warmer faster than white snow, and melts faster. Not only does this contribute to torquing the wildflower blooms, but the broken-up cryptobiotic crust releases a fungus into the air that normally lives in the soil. It gets into the lungs of susceptible people and causes an infection called valley fever.

"When you have a dust storm the spores get kicked up into the air and into your lungs," Inouye said. "A lot of prisons in California and Arizona are built in areas susceptible to dust storms," making the issue one of social as well as environmental justice.

I'll Check My Schedule

Mesmerized by desert light and the crazy abundance of wildlife in a landscape that upon first glance looks barren, I was still kind of surprised that the United States Geological Survey had decided to put its National Phenology Network (NPN) in Tucson, Arizona. After all, phenology is about blooming and dying and hibernation and migration—and yes, all these things happen here, but what about Vermont or some leafy state? In January 2014 I accompanied Jake Weltzin, the tall and exuberant executive director of the NPN, to document the day's observations on a transect he's responsible for fifteen minutes outside the city, in Pima Canyon.

"My wife very generously lets me out of the house a couple of hours every weekend so I can do this," he said, suggesting a powerful incentive for becoming a citizen scientist: respite from young children, bless them. (Eventually, they can come along. My son is a much better citizen scientist than I am.) The NPN is a program aiming to document what is blooming when, who is coming and going out of hibernation or migrating through and when, and indeed how critters are interacting all across the United States, and how all these goings-on are impacted by precipitation and temperature changes. This is the biotic world meets the abiotic world. This is the biological carbon cycle meets the atmospheric carbon cycle meets the geological carbon cycle, all moderated by time, by the turning of the earth on its axis, the resulting angles and distances to the sun creating seasons. It's all one system, and it's changing.

"We're a consortium of individuals and organizations that are getting together to understand this pattern of life cycle events for plants and animals on a national scale," Weltzin told me. "What is the timing of flowering, or leafing, in the deciduous forest? Or the timing of salmon runs? We're trying to get a handle on the patterns of plant and animal activity."

The NPN is vastly expanding the historically established good habit of taking note of what's going on outside. Weltzin told me, "The oldest phenology records in the United States are probably the Lewis and Clark data set. Lewis and Clark were charged by Jefferson to go into the West to explore, not to just open up new frontiers but to record what they saw as amateur naturalists. Jefferson told them to keep track of all manner of timing—of seeds and fruits being produced, leaves coming on, fish in the stream. They also took some stream-temperature data and kept some track of stream breakup dates." The US Geological Survey, which hosts NPN, has monitored stream temperatures and flows since 1880. "At that time, water across the US was critical, like it is now. It was an agrarian society, so people understood that the relationship between climate and weather—between the physical and the biological worlds—is the basis of life. We're beginning to relearn that."

He went on, "NPN happened here in Tucson when a group of researchers recognized that if you want to understand the impacts of climate change on natural ecological systems, you don't go to a handful of locations and measure things. You need to distribute a network, on a wall-to-wall scale."

"Wow," I remarked. Weltzin is a plant ecologist by training and formerly a tenured professor at the University of Tennessee. He has a big job now but still it's quite a big deal to walk away from the academic safety net. When he got offered the NPN position, Weltzin said, "I had started to see the potential for doing science on a national scale, much more broadly than my little research plots. It's not every day you get to start a brand-new science and monitoring program."

Weltzin was essentially charged with designing and implementing a framework "to create a whole national structure where essentially everybody is using the same protocol for data collection." This isn't easy. "Even though I'm an ecologist and most comfortable counting little seedlings of oaks as they come out of the ground from acorns—now I'm an information manager." Weltzin oversees Nature's Notebook, which is the platform by which you and I can join in and participate in taking our nation's nature pulse. "It exists online, and anybody can do it. Citizen scientists learn the protocols, how to find and identify a plant. We teach you essentially how to sort of adopt that plant, track it through time, and report your data," Weltzin told me. Of his own transect in Pima Canyon he said, "It's like my Zen. I've been up on it about two hundred times now and I'm pretty intimate with the setting."

Weltzin pointed out one of his favorites to me. "Limberbush," he said. He zeroed in on a rooted bouquet of soft flowers, rounded like the hat of a churchgoing Southern lady. "I mean brittlebush," he corrected himself. Sounded like a yoga pose waiting to happen here in the desert. Limberbush has long, arching branches and lives nearby. "What's so interesting about brittlebush is that the timing of the organism is very tightly coupled to weather conditions. Climate is just the average of weather at a given location—so weather through time adds up to climate."[21] I like his explanation because we talk about climate change all the time and grapple with the fact that it is not the feature of a single day or week or even year in time that indicates such, but a long, ongoing process embedded in history. "When it gets warm and there's sufficient soil moisture, the brittlebush leaves will soon come out, followed by the flowers. But if it's a cold year or if there's a frost, the leaves just drop off and there are no flowers."

Some plants do as the brittlebush and play dead, and others just die, "slowly or quickly, mechanism unknown," said Weltzin. Standing with him on his transect and pondering his hard-won observations, I was put in mind of what it's like to wait for a baby to be born, and also to attend to a loved one's deathbed. The mystery of life's incoming and outgoing breath was here moderated by the transpiration and respiration of plant life, keeping this desert alive and also keeping every place on terrestrial Earth alive. The desert around us was like a dry, warm blanket, winding cloth for the deceased, swaddling for the newborn.

One mechanism for killing other plant life that is known all too well has been the invasion of the Sonoran Desert by buffelgrass, introduced to the Southwest in the 1940s as forage and for erosion control. Buffelgrass spreads easily and does well in subdeserts like the one right here. It produces a lot of aboveground fuel, basically dead dry leaves that burn very hot. But the desert here and its historical inhabitants are not adapted to fire. All around us in Pima Canyon, it's easy to see how far apart the plants are, and the landscape is dotted with those iconic saguaro cactuses, arms reaching toward heaven, some of them hundreds of years old.

"What you see here is rock and dirt," Weltzin pointed out. "We do have lightning strikes but fire doesn't carry, because there isn't much to burn."[22] Buffelgrass is fated to burn and burn big. When it does, the fire will kill the cactuses, which will not resprout afterward. I'm thinking I've got to take a family vacation here soon so the kids can see the Sonoran Desert before it no longer exists. Satan's travel agency should specialize in trips to the iconic places on Earth that are disappearing as we speak. The intricate and dense variety of species that make up the desert will eventually and perhaps soon be reduced to a desert grassland monoculture, which will burn regularly—also imperiling *Homo sapiens* and their property. The buffelgrass itself will come back just fine.

Although he is a can-do guy, Weltzin was depressed about the buffelgrass. The summer of 2014 would turn out to be full of rains that would speed its growth.

"The hillsides are lost," he told me when I saw him again early in 2015. "It's mind-blowing to watch a system change so radically and thoroughly but also cryptically, like a plague. We will hit a threshold sometime, and the hillside and all on it will burn. It will be amazing to watch, and devastating."

Buffelgrass seems to sneak up under Weltzin's feet here. Another way to observe its steady spread is through satellite imagery. Weltzin and his colleagues have in fact taken an incredibly important step in the development of citizen science by combining buffelgrass imagery from MODIS, one of the US government satellites that take pictures of the earth 24/7, with citizen-collected data from Nature's Notebook. The MODIS imagery looks as if big splotches of green paint are being dropped on Arizona and environs as the buffelgrass "greens up," or leafs out. Information on the timing of the green-up and contemporary temperature and precipitation levels is helpful to those trying to eradicate buffelgrass, since the herbicide most commonly used to combat the invasive is glyphosate. This is a "contact" herbicide sprayed on the leaves and then transferred to the roots of the plant via photosynthesis (thus the plant needs to be "greening up" for the herbicide to be effective). Analysis of the pattern of buffelgrass green-up over time shows that it responds more quickly to rain than do native plants, suggesting an "optimal treatment window."

However, "when we look at the MODIS images," Weltzin told me, "one cannot distinguish between all the various plant species on the ground. Buffelgrass and brittlebush are both 'green' at the scale of two hundred fifty meters." Ground-based observations made by citizen scientists help "calibrate or differentiate the incoming signal. The tricky part is discerning signals across scales." A green plant looks like a little dot on a canvas such as an impressionist might dab, but in the satellite pictures represents about a quarter of an acre of ground. That's a big piece of land. The smaller-scale citizen observations both confirm the satellite images and help parse out the contributions of buffelgrass greenness to the overall signal.

Weltzin told me he and his colleagues have some other tricks up their sleeve for searching out the evil weed. "We are starting a new project using hyperspectral sensing and LiDAR combined, mounted on drones that will fly over the habitat looking for buffelgrass based on its structure and color." Hyperspectral sensing divides visible light into more finely tuned gradients than are discernable by the human eye. LiDAR, short for "light detection and ranging," allows analysis of reflected light to create a three-dimensional image of plants and ground. Crazy! We can see into the structures of nature like Van Gogh. "We may never 'get' it, but we may be able to target some set-aside zones—for example, in parks, and protect those. Maybe."

What the Lilacs Know

In the late 1950s, citizen scientists joined in making note of when lilacs and honeysuckles came into bloom each year and put that data on postcards in the mail. This historical phenological observation network, since incorporated by NPN, was established to help figure out weather patterns and inform crop management. Eventually a few thousand contributing groups included weather service observers, scientists, technicians at agricultural stations, and garden club members. A professor at Montana State University, Joseph Caprio, started up a western program in 1956 to watch lilacs, and similar projects were later established in the eastern and central states. These projects came and went and were revived by others so that we actually have a pretty good picture of historical lilac phenology between 1956 and today. One scientist wrestling with the resulting data is using it to try to discern if lilacs are picking up a signal from the environment and perhaps taking a cue to begin adapting to a new climate.

"I don't know what the lilacs know!" Toby Ault, a Cornell University climatologist told me. "Our planet has gone through these amazing changes. New York was under two miles of ice twenty-four million years ago. Lilacs were around. And they were able to cope with a major shift in the climate spectrum, around one hundred twenty thousand years ago. Any organism that responds to the transition from winter to spring"—as plants do by leafing out when the weather starts to warm up—"picks up information from the atmosphere that is relevant to them. From the soils below to the air above, the amount of sunlight, length of days, all of the indicators, all of the qualitative features that define spring come into relevance to a plant coming out of dormancy."

Plants have to detect when it's warm enough to get the show on the road to maximize the good growing season but avoid getting zapped by frost. "They have to make a biological choice," Ault said. "If it was simple they would leaf out the same day every year, because the length of days does not change from year to year. We know it's not random and we know it's not like popcorn with species going off; there are large coherent spatial patterns that synchronize events. This large-scale synchrony is a footprint of the atmospheric conditions, the whole climate."

Ault is a citizen science enthusiast and enjoins Nature's Notebook participants to plant lilacs and monitor them for him. "The dialog goes both ways," he said. "Citizen science is a research tool and an apparatus for monitoring, but it is also a way to cultivate a scientifically oriented society." He told me that getting at the plasticity, or ability of species to change in response to climate conditions, "requires interdisciplinary scientists and huge networks of citizen scientists." What we will do with information about how plants respond to information we have yet to discern also remains to be seen, but having a deeper handle on the processes at work here will undoubtedly influence all kinds of social decision-making, including management of the nature we have left.

"We want a continent with more than a handful of species on it," he said. "This isn't only a story about lilacs. There are thousands of other species we can learn the signals for. That's why we need twenty million citizen scientists."

Big Data

Toby Ault and Jake Weltzin are dealing with patterns across gigantic scales perceived by way of statistics and computing power. It is potentially much more generalizable than experimental, hypothesis-driven science, which necessitates human-constrained parameters that are time- and place-specific—ultimately big-data ecological science wants to work with all time and the whole globe. Hypotheses themselves are embedded with subjective layers. Arguably so is the decision to monitor lilacs, but measured against the hard facts of date, time, and temperature, the pattern that emerges essentially speaks for itself. Memorably I once asked David Ackerly, who heads up the Berkeley Initiative in Global Change Biology as well as the Terrestrial Biodiversity Climate Change Collaborative, why he likes statistics so much. He looked at me with a funny expression and I thought I must have just asked the stupidest question in the world. (As Joseph Campbell says, the job of the journalist is to educate yourself in public.) But Ackerly was thinking. "Statistics," he answered, "let me know the story I'm able to tell." It is somewhat ironic that the citizen scientist, arguably the most subjective, most "I"-based, observational locus you can find, is the necessary complement to this sort of science.

CHAPTER NINE

Innocence and Experience

By the time I graduated from college and headed into my own literary life, my father was deeply ensconced back in the advertising world from which he had had a ten-year hiatus. We went out for simple lunches or dinners every couple of weeks in New York City. He paid—I made so little money, these meals provided me with a serious nutrient boost, as a conservation biologist might put it. I reported to him the highs but not all of the lows of my first job as assistant to the editor in chief at *Esquire* magazine. Eventually, grasping for meaningful copy beyond the typing pool, I was handed the monthly column of a European playboy to edit.

Esquire in those days was revered among the literary establishment partly for publishing what was called "new journalism," in which the reporter included himself as a subject of the story (very few women broke into this world)—the observer as participant. My boss was a young businessman who had bought the magazine. While respectful, he did not feel beholden to the old *Esquire*. What I absorbed from working with him was that he wanted to empower people as he had empowered himself, to realize their full inner potential and to create their own lives. He recruited "New Age" California writers to the pages. The old guard who still worked there were not amused. Neither camp was particularly concerned about my European playboy (which was why I got to edit him), who wrote with flat ease about tennis, drinking, and beautiful women. Then he went on to beautiful women, drinking, and tennis. Editing him was like mowing a lawn.

I was in my very early twenties and he was around forty. Here's how my interactions with him went. Me: "X, I'm afraid I'm going to have to cut your copy—by a lot." X: "Fine. Mary Ellen, will you sleep with me?" Me: "No. Do you want to review it before it goes to the copy editor?" X: "Please, if you would sleep with me . . . it would be such a very good idea, I want to sleep with you." Me: "Do you want to review the text before or after the copy edit?" X: "No. Please, go to bed with me." Sometimes he modulated his voice, rolling the tone. Mostly not. X: "Please . . . if you would only . . ." Me: "Talk to you next month!" Once I found myself in a taxicab with him. I was leaving a party and he hopped in. We drove downtown and he kept up his pleading. I realized I did not want to be standing next to him at the door to my apartment building. At a stoplight I bid him good night and stepped out of the cab. Our professional relationship was not impacted.

There was a lot of typing of manuscripts I was not otherwise involved with, lots of picking up dry cleaning and fetching lunch; there was a garden variety of casual sexual harassment, and this was the literary life. Then one day I got a phone call from a woman I had gone to college with but didn't know well. She had tickets to a lecture series at the New School she couldn't use, and I was the only person she could think of who might be interested. The tickets were for three separate lectures by Joseph Campbell. I was impressed that the price of the tickets nearly equaled my monthly rent. If I had had to pay for them myself, I wouldn't have been able to eat for several weeks. What a biologist might call an "extra resource allocation" could be called in Joseph Campbell's term a boon received as an *aide de journey*.

Campbell's *Hero with a Thousand Faces* was by then a cult classic but I hadn't read it. A few years later journalist Bill Moyers produced a PBS series on Campbell, *The Power of Myth*, and it made Campbell world famous. George Lucas credited Campbell with inspiring his *Star Wars* movies and thus fueled a trend that continues today for screenplays to be written and evaluated according to the concept of "story structure" as articulated by Campbell's hero myth. If you would like to see some nifty diagrams of the hero's journey, Google "Joseph Campbell and Marketing." You will see that his story structure is also currently in wide use as a key to selling products, oneself, or both. A favorite Campbell aphorism in use by best-selling author Paulo Coelho and to herald a Pinterest board is: "Follow your bliss and the universe will open doors for you where there were only walls."[1] From the commercial to the esoteric, the stories we tell today, the stories of our time, if you will, are in large part shaped by Joseph Campbell's storytelling.

Photos of Campbell and tales of his youth depict a very handsome young man, though at the time I saw him he was in his late seventies. He was tall and robust, but somewhat indistinct, like any businessman in a suit on his way to Grand Central Station. The lecture hall at the New School was packed. Campbell stood on the stage and talked. That's all it was—a man and a voice. The audience was riveted, including me.

Campbell's main point was that life is not about telling ourselves a story, it is about *living* a story. He exhorted his listeners to follow their "bliss" into a deeper plane of authenticity and experience. As he later told Bill Moyers, "Most people living in that realm of what might be called occasional concerns have the capacity that is waiting to be awakened to move to this other field." He told Moyers about teaching at a boys' prep school, where the students would consult with him about their career paths. (This would be the Canterbury School in Connecticut, where Campbell was a student before he was a teacher; he matriculated alongside the writer John McPhee.) "A boy would come to me and ask, 'Do you think I can do this? Do you think I can do that? Do you think I can be a writer?'" Campbell responded that if the boy could withstand ten years of obscurity, the answer was yes. "Then Dad would come along and say, 'No, you ought to study law because there is more money in that, you know.' Now, that is the rim of the wheel, not the hub, not following your bliss. Are you going to think of fortune, or are you going to think of your bliss?" At one point at the New School he recounted a Sarah Lawrence College student's response to his hero's journey lesson (he taught there for decades). "'Professor Campbell,' she said, 'I understand it. I get what you are saying and I know how to attain wisdom.' 'Terrific,' I replied, 'all you've missed out on is your life.'" His delivery was so spontaneous I was truly surprised, reading him later, that he repeated this anecdote regularly in slightly varied forms. At the time, like many in the audience, I felt he was speaking directly to me.

Campbell later located the decisive moment in his own heroic journey to a period of eight months during the Depression, in which he hung out with John Steinbeck and especially Ed Ricketts in Pacific Grove on California's Monterey Peninsula. In one way of looking at it, the friendship between Steinbeck and Ricketts was a long conversation about art, science, and what it means to be human. In 1932 Campbell entered the conversation, influencing it and being influenced by it. "A beautiful time," Campbell called it. "I'm coasting along, trying to find where I am. . . . Ed Ricketts was an intertidal biologist. We'd go out and collect hundreds of starfish, sea cucumbers, things like that, between high and low tides, furnishing animals for biology classes and schools."[2] Elsewhere, he recalled, "It was Ed who was especially important to me. . . . From our long talks about biology, I eventually came up with one of my basic viewpoints: that myth is a function of biology. It's a manifestation of the human imagination which is stirred by the energies of the organs of the body. . . . In other words, myth is as fundamental to us as our capacity to speak and think and dream."[3] As Campbell later said to Jungian analyst Fraser Boa, "Ed Ricketts was the heart of it all."[4]

The Journey Begins

The story of Ed Ricketts is a case study of the hero's journey eventually articulated by Joseph Campbell. Whatever else he did or didn't do, Ricketts followed his bliss. And plenty of others to this day follow their own, in his footsteps. As evolutionary biology continually references Darwin, a subset of literary seekers, often aspiring to integrate science with its sister humanities, find in Ed Ricketts a powerful touchstone. On several expeditions, including the one to Sitka, Alaska, with Campbell and another to the Gulf of California in Mexico with Steinbeck, Ricketts collected specimens that inform how we understand change over time. But Ricketts was equally interested in subjective questions of meaning and purpose as he was in objective data points, and in this he is of significant relevance to us today. Science alone is not going to save nature; context is everything.

In full-on searching-young-man character, Ricketts made three departures from traditional education (he never received a college degree). Starting at Illinois State Normal University in 1915, a year later he was doing stints as an accountant in El Paso, Texas, as a surveyor in New Mexico, and as a Medical Corps technician at Camp Grant in Illinois. Next he matriculated at the University of Chicago but, initiating a pattern he would repeat, left there to escape the consequences of a dalliance with a married woman,[5] and in 1916 commenced a *Wanderjahr* through Indiana, Kentucky, North Carolina, and Georgia, mostly on foot. In 1925 he penned a charming recollection of his walkabout in the South, "Vagabonding Through Dixie," published in *Travel*.[6] (Ricketts' walkabout was also described in Steinbeck's 1945 book, *Cannery Row*.) Ricketts, who was visibly armed, recounts being taken for law enforcement more than once. "As I entered a little country church one Sunday evening, I overheard a youngster telling his companion, in a hoarse whisper, 'Thar goes a Rev'noo Off'cer.'"

Coming across John Muir's *Thousand-Mile Walk to the Gulf* in a Savannah, Georgia, library, Ricketts decided to follow the legendary naturalist's suit and "whenever it was convenient, I spent the night in a 'City of the Dead.' There was always good shelter and a level place to pitch my tent." As Muir writes of the practice: "On rising [one morning] I found that my head had been resting on a grave, and though my sleep had not been quite so sound as that of the person below, I arose refreshed."[7]

John Muir is another citizen science prototype and merits a substantial aside. He made at least one highly significant contribution to science, in 1871 establishing (without an advanced degree) that Yosemite Valley was created by glaciation. This was contrary to official consensus at the time. He resisted the category but was thought of by professionals as one of California's foundational scientists.[8] Muir advocated for direct experience. "No scientific book in the world can tell me how this Yosemite granite is put together or how it has been taken down. Patient observation and constant brooding above the rocks, lying upon them for years as the ice did, is the way to arrive at the truths which are graven so lavishly upon them."[9] At the same time, Muir also reprised the practice of devout Europeans by interpreting nature in terms of physical texts. Writing for the *New York Tribune* in 1871 (his first publication for pay, $200), Muir compared Yosemite glaciers to a weather-beaten book: "And though all were more or less stained and torn, whole chapters were easily readable. In this condition is the great open book of Yosemite glaciers today; its granite pages have been torn and blurred by the same storms that wasted the castaway book."[10]

In 1921 Ricketts returned to the University of Chicago, where he spent the next two years. One of his professors made a deep, permanent mark. Ricketts took hard to the ideas of zoologist Warder Clyde Allee, which made their way into Steinbeck's work as well. Allee was a proponent of "the organismic-community concept" and emphasized that the individual doesn't count for much in ecology, but the group does. A practicing Quaker, Allee sought to integrate ethics and science and argued that animals benefit from living in cooperation. He asserted that "the two great natural principles of struggle for existence and of cooperation are not wholly in opposition, but… each may have reacted upon the other in determining the trend of animal evolution."[11] Allee expressly looked for biological grounds upon which to argue the benefits and naturalness of cooperative human societies, posing the idea in counterbalance to the every-organism-for-itself way of looking at Darwin's definition of natural selection. Allee melded observations of tide pool interactions into a social vision,[12] and they are reflected back in Ricketts' lifelong musings on the nature of tide pools and everything else.

From 1915 to 1921, Allee conducted an intertidal survey at Woods Hole in Massachusetts along the lines that Pete Raimondi and Melissa Miner are doing on the West Coast today, and for which at the time there was little precedence. As discussed earlier, Joseph Grinnell was famously (now) in the process of surveying terrestrial California in a highly replicable way. The utility and even necessity of surveying like this was not generally known at the time and Grinnell was prescient. Ricketts probably learned the concept of the biological survey from Allee, but neither of them used their data in the highly quantitative manner Grinnell pioneered and which present-day scientists pursue with greater accuracy every year. Allee was after a behavioral result, or one might even call it a spiritual result—he sought to show that cooperative behavior in "lower" animals is characteristic of living matter itself.[13] This is not how today's biologists would frame their questions in a survey. Joseph Campbell would have been right at home with this sort of biological and aspirational synthesis. Ricketts revered Allee and considered his the "last word," and according to his friend Jack Calvin, Ricketts and other Allee students "got a holy look in their eyes at the mention of his name."[14]

While he attended classes, Ricketts worked at Anco Biological Supplies, part of a fishing-tackle supply company, an experience that likely prepared him for the business model on which he built Pacific Biological Laboratories with Albert E. Galigher in Pacific Grove, California. Ricketts arrived there in 1923 with his wife, Nan (Anna Barbara Maker), and young family in tow. Galigher would soon step out of the business, and Ricketts soldiered on alone.[15] "What Ed did for a living," as described by

Steinbeck biographer Jackson Benson, "was to collect, prepare, and ship animals to schools to be used for exhibition, experiment, and dissection in high school and college biology and zoology classes."[16] He gathered most of his specimens from the Monterey intertidal, but also took collecting expeditions to far-flung parts of the West Coast.

The year he published "Vagabonding Through Dixie," 1925, Ricketts also published a catalog of specimens for sale from Pacific Biological Laboratories. Thus he prefigured the interest in and expression of two modes of experience that later he and Steinbeck would attempt to fuse in *The Log from the Sea of Cortez*, the travelogue and specimen list memorializing their expedition around the Baja Peninsula. Ricketts' foreword to the lab catalog begins Big Picture: "Monterey Bay is the fusion point of faunas from the North and South, and the ranges of a number of characteristic species of both regions overlap in these waters. . . . One can find almost any combination of rocky coast, sand beach, or mud flat within a few miles. Rich pelagic hordes approach the shore." He concluded the foreword with a most unusual caveat for the time. "It should be borne in mind, (and this applies especially to local marine forms), that we must, above all else, avoid depleting the region by over collecting. . . . Monterey Bay is probably richer in individuals and species than any other region of like size in the United States, and it would be unfortunate if such a situation were to arise here."[17]

Revisions

The mind meld that would form around Ed Ricketts in the thirties began with the arrival of Jack Calvin to Carmel in 1927.[18] Both bookish and adventurous, the teenaged Calvin had traveled by square-rigger to Bristol Bay in Alaska to work the canneries and to fish the Nushagak River. He wrote two young adult novels based partly on these experiences, *Square-Rigged* (1929) and *Fisherman 28* (1930). "They're both out of print," Lisa Busch, executive director of Alaska's Sitka Sound Science Center, told me, "and we're trying to get a commercial publisher interested in reissuing them. They're fantastic seafaring adventure tales."

Busch is among those who believe Jack Calvin deserves a higher place on the marquee for his part in writing the book Ed Ricketts often gets total credit for producing. Busch is a co-investigator on a project funded by the North Pacific Research Board, named for the title in question: "Between Pacific Tides: Revisiting Historical Surveys of Sitka Through Ricketts, Calvin and Ahlgren."[19] It's a model citizen science project, providing multiple tiers for involvement, on the one hand delivering hard-core science and, on the other hand, in the words of Busch, "delving into the whole history, the whole time-and-place-and-personality aspect from which scientific thinking emerges."

Jack Calvin was married to Sasha Kashevaroff, one of six daughters of a Russian Orthodox priest from Alaska. Sasha's sister Natalia, or Tal, married Ritchie Lovejoy, a local Monterey artist who contributed most of the original drawings for *Between Pacific Tides*. Calvin, Lovejoy, and the Kashevaroff sisters were, with Steinbeck, among Ricketts' closest friends. (Another Kashevaroff sister played a key role in his life as well, to be discussed anon.) The swirling overlap of friends, bylines, and expedition companions attests to the open-source, collaborative nature of Ricketts' relationships.

On the science side, the project established the historical survey sites where Ricketts and Calvin combed the tide pools at an important juncture in the development of *Between Pacific Tides*, which was an innovation in its day, since Ricketts organized his subject by where you find creatures in the various intertidal zones—he intuitively reflected Joseph Grinnell's nuanced concept of the niche. Ricketts collected specimens in Sitka with Calvin and Campbell, and these now represent a baseline point of comparison with specimens documented from the same location, decades later. Today's Sitka project is structured to feed rigorously obtained data into a scientific database. "The Sitka people wanted to take some of Ricketts' and Calvin's old journal documenting what they had looked at, their classic intertidal surveys, and to do a comparative study between then and now," Pete Raimondi explained. "They asked if we were interested in re-setting up their sites." Raimondi was thrilled with the idea, partly because a rocky intertidal monitoring site in Sitka would appreciably extend the range of his Pacific coast monitoring program. With Miner and the Sitka folks, Raimondi set up resurveying sites in Sitka in 2011 and has monitored them every year since.

On the cultural side, the resurveying of Sitka has to do not just with tracking ocean changes and adding data points to the larger coastal project, but with deepening the appreciation of Sitka as a nexus of art, science, and history much in the way we appreciate Monterey. "Calvin and Ricketts made a seminal contribution to marine biology here," Busch told me, "but Sitka itself is not fully aware of its role in the history of science."

Calvin and his wife, Sasha, were frequent tide pool companions of Ricketts in Monterey, and Calvin especially encouraged Ricketts to integrate his collecting observations with scientific context and to publish a beachcomber's guide. Ricketts took up the idea and enlisted Calvin to help. What Ricketts wanted to produce though was vastly more ambitious in scope. Ricketts scholar Katharine Rodger told me that Calvin undoubtedly helped with the first drafts of the book. Ricketts was "ultimately an open-source kind of a guy, and freely turned his work over to other people." Rodger added, however, "The book was Ricketts' by design and organization. It was born out of his mind. Ricketts was the one thinking about ecological organization and the environment."

Calvin was a good writer, full of humor and unexpected detail. A 1933 *National Geographic Magazine* story about his and Sasha's honeymoon trip, "'Nakwasina' Goes North: A Man, a Woman, and a Pup Cruise from Tacoma to Juneau in a 17-Foot Canoe" is the stuff of today's Patagonia clothing company catalogs.[20] A central grist of the journey was the canoe itself: "She was never manufactured, that canoe; she was created . . . built with the painstaking care that goes into a fine violin." Jack, Sasha, and Kayo traversed one thousand miles by sail and by paddle, making their way up a coastline of lichen-like intricacy.

Among the Calvins' adventures were several attempts on the part of commercial fishermen to rescue them; when assured that the little family of travelers in their slender bark were in fact doing this on purpose, there was often "consternation aboard the gasboat; sometimes polite disbelief." Sometimes they paddled after dark, and in an episode that Calvin refers to as a "tragi-comedy," Kayo awakened them one night. Calvin writes: "I hauled an arm out of my warm sleeping bag and plunged it down into six inches of cold salt water. We were floating, except for our shoulders!" Hurriedly pulling themselves and their belongings onto land, in the dark they fell asleep on the closest hard, flat spot, to be awoken again by "two great moons, shining side by side," and a voice demanding why they were sleeping in the middle of the highway. Calvin left Monterey in 1932, eventually to found Alaska's oldest conservation group, the Sitka Conservation Society, which among other nature protections today enjoins citizen scientists to help monitor human impacts on the Tongass National Forest.

Stanford University Press contracted to publish *Between Pacific Tides* in 1936 but did not actually do so until 1939.

Contemporary scientists reviewing the manuscript objected to what we now recognize as its revolutionary approach in presenting species as inextricably associated with the places in which they are found. The book bore both Calvin's and Ricketts' bylines. Subsequent additions have added a panoply of contributing names including Steinbeck's, reflecting new edits, new forewords, all contributed by those eager, it would seem, to share a literary moment with Ricketts. Not only does a push-me-pull-you ensue around the question of authorship of *Between Pacific Tides*, so does a corollary disputatiousness arise around the Ricketts-Steinbeck relationship. There are those who resent Steinbeck's portrayal of his friend in *Cannery Row*, basically finding it reduces him to a shallow sybarite, when the actual Ed Ricketts was pleasure-loving but also dead serious and seriously accomplished. Others take issue with the authorship of *Sea of Cortez*, though Steinbeck himself defended Ricketts' contribution. There's something compelling about owning a particular memory of a particular person: it restores to us our own personal past, otherwise to be washed away by the incoming tide.

Party Animal Aggregations

Monterey in the 1930s represented a special data point on the space-time continuum. John Steinbeck and his first wife, Carol, waited out the Depression in his father's cottage in Pacific Grove. In 1929, Jack Calvin was living in Carmel. Steinbeck and Calvin had known each other at Stanford and now reconnected.[21] (Steinbeck never graduated from Stanford, but no one has ever held the lack of a degree against a novelist.) It was at one of his frequent dental visits that Steinbeck is reported to have met Ed Ricketts,[22] though other accounts put the fateful meeting at Jack Calvin's house.[23] Carol described Calvin's house as "very friendly," a spirit that thrived in the general Depression-era poverty experienced by all. Carol recounted the simplicity of making a party at the time: "You would gather in a ring, sitting on the floor and drink a jug of wine." Carol also noted Ricketts' neatly trimmed beard and remembered him dancing around the fire at the Calvins'; she remarked that people took him for a "Christ-like father confessor," though her description summons more of a devil.[24]

However they met, Steinbeck and Ricketts began one of the most famous of literary friendships; Ricketts was thirty-three and Steinbeck twenty-eight. They were something of an odd couple; the famously ornery Steinbeck seems to have been soothed by the warm, expansive Ricketts. In addition to the intertidal invertebrates, Ricketts focused on a broad array of enthusiasms, including music, poetry, literature, and philosophy. Steinbeck shared these interests and was also primed to talk biology.

Years later Steinbeck recalled being strongly influenced by the work of William Emerson Ritter, a major marine scientist whose ideas shaped the agenda at Stanford University's Hopkins Marine Station when Steinbeck took classes there in the early 1920s. Like Allee, Ritter emphasized a holistic view of ecology and extended his ideas into philosophy and social systems. Ritter articulated a superorganism idea, asserting that nature in all its separate parts and in its aggregate constitutes "one gigantic whole," the parts of which are "mutually constitutive of each other. Structurally, functionally, and generatively, they are reciprocals of one another."[25] In the long dialog that was the Steinbeck-Ricketts friendship, the ideas of Allee and Ritter flowed continuously. Where contemporary science tends to slice and dice nature ever more finely to isolate what is going on in it, Allee and Ritter represent a time in the history of ecological thought when the impulse was more toward integration. Science has bounced back and forth between these dichotomous endpoints; hearkening to our theme, between data and lore. Steinbeck and Ricketts loved the lore, but in their own way, they wanted to base their stories on data.

In terms of the story structure Joseph Campbell would eventually articulate, Ricketts modeled the heroic figure in many Steinbeck novels, especially *Cannery Row* and *Sweet Thursday*, both of which featured "Doc" as the central wise character, standing apart but not uninvolved with the melee around him. *Cannery Row* and the man behind the simpler outline have inspired generations now of professional and amateur marine biologists. Steinbeck structured the novel as a series of vignettes that unfold concurrently, like detailed intertidal zones revealed at low tide. He imbued important characters in nearly all his books with the wise forbearance of his good friend. Jim Casy in *The Grapes of Wrath* utters the very Ricketts-like aphorism, "There ain't no sin and there ain't no virtue. There's just stuff people do."

As Ricketts devoted his daily attentions to the species organization in tidal pools, in the early 1930s Steinbeck found a similar coherence among people, which he called "the phalanx." Steinbeck's phalanx bears relationship to the collective unconscious as limned by Carl Jung, a shared memory of, in Steinbeck's words, "a time when the moon was close, when the tides were terrific." The individual does not have this memory, but the whole group of *Homo sapiens*, larger than the sum of its parts, does. "Religion is a phalanx emotion and this was so clearly understood by the church fathers that they said the holy ghost would come when *two or three were gathered together.* [Steinbeck's emphasis]" He described the differentiation of self from surroundings as "an emotion." Steinbeck scholar Susan Shillinglaw calls out this discernment of single and group man as a Steinbeck preoccupation, emerging in virtually everything he wrote.[26]

Ricketts' own writings include two essays that attempt to penetrate biological thinking with a philosophical tincture, "The Philosophy of Breaking Through" and "Essay on Non-Teleological Thinking." Like Steinbeck, Ricketts yearned to describe an emergent quality, a "new thing growing synergistically out of component parts." Ricketts saw this new thing as transcending dualities of all kinds, "form and function, matter and energy, material and spiritual," in Ricketts' words, which are in the end vehicles purposed "toward an integrated growth."[27] Ricketts repeatedly invoked the holism of Allee and Ritter, explaining in 1940 that "the whole consists of the animal or the community in its environment, the notion of relation being significant."

Ricketts' worldview is also key to Steinbeck's ecological vision. Both Ricketts and Steinbeck sought a holistic comprehension of the natural world. Humans take a place in this integrated world but do not dominate or determine it. Our contemporary woes are orders of magnitude more dire than those they recognized, and this fundamental impulse to resize the human form on the landscape has become imperative.

Ricketts inspired passionate company while alive and now in retrospect. Katharine Rodger, who edited his letters and assorted other writing projects, laughed telling me that "most of us working on Ricketts are nut jobs. I am approached by so many people who are like..." She laughed again, her thought trailing off.

"Like a sketchy assortment from *Cannery Row*?" I asked.

"Exactly," she answered. Even for a generous reader not inclined to judge too harshly, Ricketts is hard to read. "I've talked to physicists and I've talked to philosophers, and it's pretty clear Ricketts did not really understand the 'unified field hypothesis' or non-teleological thinking," Rodger said. As Shillinglaw observed, Ricketts was looking to encapsulate his mode of looking "doubly open"—in his wide embrace, the objective, the non-objective, and everything in between, which he expansively called the "divine geometry."[28]

Teleology describes the idea that there are final causes in nature that determine how things come to be. Darwin, for example, is the ultimate non-teleological thinker, since natural selection by definition does not have a goal. "I don't think his value is in scientifically quantifying these things," Rodger added. "He was about trying to figure life out. Without sounding corny, he was looking for the answer to the universe." He found hints in Bach, Whitman, and personal interactions. Although he described Ricketts' mind as "without horizons," Steinbeck also once commented that a part of Ricketts was "walled off." Accessible and inaccessible at the same time can indeed be a recipe for charisma, and this indistinctness may actually be at the heart of Ricketts' appeal. It also suggests an emotional dimension to his quest for "breaking through."

I do love this anecdote about Ricketts' response to a recital by John Cage at one of the Carmel parties. Cage arrived early because he "wanted to fix the piano," which entailed placing two large knives, a piece of wood, and empty tin cans among the strings inside it. The audience of classical music lovers expected something other than what they got, and reacted hard. "Older people looked bewildered, almost as if in agony. Younger people more familiar with jazz, however, seemed elated and on the verge of applauding." A classical music aficionado, Ricketts sat on the hearth, "his head in his hands." He did not declare yay or nay in the heated discussion that followed Cage's playing, but said, "What a future that boy must have to get a reaction of such intensity and magnitude."[29]

Existential Forensics

Pacific Grove in the 1930s was something like East Hampton

in the 1970s. The vibe was festive, with a higher purpose, and among the parties and the fun, lasting contributions were being created by foundational artists. One partygoer at Ricketts' lab remembers it as a place where the spirit was "'let's dig, let's find out'–a spirit which was in large part generated by the presence of Ed."[30] This communal coherence is something today's "Ed Heads" refer to with nostalgia and would like to re-create. I had forgotten that my own childhood was imbued with the same sort of heightened tone until it was grieved by the community at my father's wake.

"No food? No wine? No water?" I asked my mother, disbelieving, and wondering why anyone would want to come to such an event, when the day after it there would be a mass and then a reception that would furnish forth. "This is the way they do it," she whispered as I walked with her into the dusk-darkened funeral home to make the macabre delivery of my father's favorite turtleneck in which he would take his final rest. I asked these questions again of the employee behind the desk. He shook his head and did not look me in the eye. Peering past him I watched someone else's wake going on. An old man's head reposed in an open coffin. About twenty people milled around talking to each other. Nobody had a glass or a plate. Okay, I get it, I thought. My father's wake will be attended by the most dutiful of friends, and then we'll see everyone else at church and at the party afterward.

The next evening my mother, my siblings, and I stood in front of the closed casket. The funeral home took up residence in a building near the house I'd grown up in on Pantigo Road. When getting dropped off at home by friends, I'd always guided them, "It's three houses past the medical center." Perhaps this was the very space where I'd stood getting my gashed thumb bandaged up, reciting letters on an eye chart, sticking out my tongue and saying "ah." Like pretty much every other square inch of East Hampton, the place had been completely made over into something it wasn't before. For years when I visited, I spent a lot of time squinting to reconvert the evidence of my eyes, miles of gigantic mega-mansions, back to the open fields of my youth. I disrobed the fancy window displays of Main Street to re-see the plain pharmacy and the homely newspaper store.

The past lived in my mind but it was not here anymore—as Thomas Wolfe put it, you can't go home again. I'd long given up feeling sorry for myself about it. People all over the world have been and are being disrupted from their homes and, comparatively, it was not a tragedy that East Hampton abandoned itself long ago. Watching the beauty go had been painful, but this place sat on a spit of land jutting into the Atlantic. A climate change–fueled hurricane was probably online for rearranging the real estate listings. Joseph Campbell did not include this dimension of the hero's journey in his schema. *Homo sapiens* are migrants and we often leave a place because we have to.

People started to arrive. Relatives from out of town. Friends from New York who had made the three-plus-hour drive. Neighbors. Despite the earnest hugs and the glassy eyes, it started to feel more like a wedding than a funeral. It seemed that not only everybody I knew, but everybody each of my siblings and my parents knew, waited patiently in what had now formed into a line. People I had not seen or frankly even thought of for decades materialized. Fishermen, artists, bankers, restaurateurs. The younger brother of one of my siblings' close friends had somehow turned into a middle-aged man who said, "Sorry for your loss. I had a big crush on you in high school." "I had no idea," I said, wondering if I should thank him, and was instantly transported back to the cinder-block walls of East Hampton High School. The image was jarringly deepened by the next in line. "Mr. Sarlo!" I exclaimed to our militaristic principal, who had not aged an iota past a few more gray hairs in his buzz cut. He told me to address him by his first name and I said, "Okay, Mr. Sarlo." The line started to extend beyond sight.

Eventually people were turned away; friends told me later about waiting for two hours and not getting in. I had no sense of an endpoint to the bodies coming through the door and snaking around the somber, shadowed room. It was a river of individuals bearing data points of memory. It was a map in reverse; the destination, the end of this life, had been attained, and we were reviewing how we got here. I was happy to see each of them, though the processional put me in mind of Dante's *Purgatorio*. Was my father beloved? It wasn't that—people had affection for him and his outsized personality. He represented East Hampton back in the day, the artsy, community-cohered enclave where farmers and poets steamed lobsters together on the beach. A time that was gone but, for this one moment, repossessed.

"Mary Ellen." Loudly, as if to geolocate me with his voice, Leif Hope presented himself with his daughter Nissa. A pair who had aged and yet not. Hope was a landscape and portrait artist who stood with the masculine assertion of his heyday, thumbs tucked into big fists absently awaiting the paintbrush. While famous writers, including John Steinbeck, contributed to the intellectual life of the Hamptons, the nature of the place was established by visual artists. In the late decades of the twentieth century, the ghost of Jackson Pollock yet presided, the paint-splattered floors of his house open for view. Pollock famously said his art was about the rhythms in nature. Many painters made and remade realities in the ocean-saturated light. Leif Hope had come to East Hampton in their footsteps; my father had come here in their footsteps. Here was Nissa, for whom I had babysat with her brother Erling, as pretty a fortysomething as she had been a child curled on the couch, and when she put her arms around me we were both back there, carrying the freight of childhood—as Rilke avers, the place beyond which is not the future.

Like the wildflowers carefully documented by Padre Juan Crespí as the Spanish approached Santa Cruz, these individuals swayed in the suns of memory. And yet here they were again, re-emerging to make one collective salute to everything each of us had gained and lost in the nearly fifty years my father lived in East Hampton, creating it all one more time. As John Steinbeck might put it, the phalanx produced something more than the sum of its parts. Ed Ricketts would have recognized a breaking-through. And Joseph Campbell would have said that the most important moment in a hero's life is his death.

The Boon

The party in Pacific Grove was in full swing when Joseph Campbell dropped into it. Sidelined by the Depression, Campbell drove across the country in a decidedly undirected way. Years away from recognizing his calling as world mythologizer, Campbell was assiduously working at becoming a fiction writer. Landing in Pacific Grove, he quickly met and hit it off with Steinbeck and Ricketts. The eclectic conversation on their first evening together covered religion, art, fireplaces, and Los Angeles. Campbell stayed up and listened while Steinbeck read to him from a novel he had been writing for five years, *To a God Unknown*. Steinbeck's second book explores a pagan connection between man and the land disrupted by Christian doctrine. This would be catnip for Campbell, who mused: "I had the curious feeling ... as I met Steinbeck and he walked towards me. I thought I was seeing myself."[31] In Campbell's description, both men were "serious and sturdy," and were once mistaken for brothers.

Campbell fell fast and easily into the rhythm at Pacific Grove. There was target shooting at Point Lobos and collecting trips to Santa Cruz. There were deep-into-the-night discussions of essential matters, let's-get-together-and-read-a-book parties. In a story that became a cornerstone of his own creation myth, Campbell confided to Ricketts that he was at an impasse. "I said to Ed, 'I've just been saying no to life.'" Ricketts advised getting drunk and mixed up a brew with laboratory alcohol. Campbell heartily embraced his initiation: "I was twenty-eight at the time and in perfect shape, and I dragged them all under the table... until about two o'clock in the morning." One taste of the house brew quickly curtailed a police inspection of the scene. The evening included a foray to town to watch "a chap" in pursuit of the world's record for roller-skating on a flagpole, which is actually kind of hard to visualize. "It was my party to start me off in life!" Campbell exclaimed, and said Steinbeck wrote about it in *Cannery Row*.[32] Campbell might have also recognized himself in Steinbeck's *Sweet Thursday*, which features Joe Elegant, a solipsistic cook laboring over a novel called *The Pi Root of Oedipus*. This stab at Campbell was not without cause.

Campbell came to Carmel ready to experience life, having studied it assiduously. The perspective he brought to Ricketts and Steinbeck was not yet fully baked, but the ingredients gathered and mixed; in a diary from his late teens he wrote: "It seems to me that God meant man to study science," repudiating those who "scorn science as a heresy and seek all their knowledge in the Bible," and concluding: "I do not doubt that Christ's miracles were many of them a perfect application of a perfect science."[33] He began college at Dartmouth studying biology, but transferred to Columbia, where he studied English literature.[34] Campbell flirted with the subject of anthropology, and listened in on lectures by Franz Boas at Barnard College next door.[35] Given that Campbell's life work would eventually focus on world mythology, emphasizing a common humanity across race and time, one wonders that he didn't stick with Boas and help launch American anthropology. Evidently he was meant to stand alone and not as part of a collaborative tradition.

Campbell went on to get a master's degree in medieval studies—and here is one of many affinities with Steinbeck. Over his lifetime, Campbell would round up basically every cultural tradition he could get his hands on to reveal the ubiquity of the hero's journey in the historical human psyche, but the Arthurian legend he focused on in his master's thesis most comfortably fit around his own broad shoulders. It also fit around Steinbeck's shoulders. Steinbeck's last major literary effort would be a retelling of the Camelot story, which he abandoned without finishing. Campbell had helped disrupt the marriage of John and Carol Steinbeck, acting the Lancelot role with King Arthur and Guinevere, almost as if playing to a mythic script.

Where Steinbeck's sensibility was nursed in the landscape of a particular place, California, Campbell's background was more cosmopolitan. His father was a salesman for an apparel company and did just fine until the Depression hit. The family valued education and traveled together as a matter of course. On a 1924 ocean crossing to England, Campbell found himself on board a steamship with a teenaged Jiddu Krishnamurti, who would go on to become a world teacher of individual fulfillment and transformation without organized religion. Campbell brought Krishnamurti's thinking into the Steinbeck-Ricketts circle, where it was highly influential. He later explained that Krishnamurti's teaching "had to do with integrating all the faculties and bringing them to center. He used the image of the chariot drawn by the three horses of mind, body and soul. This was exactly in my line."[36]

In a little tuck of the cosmic weave, Krishnamurti would later become friends with Robinson Jeffers, whom Campbell never met but whose poetry became an instrumental source for the group convened in Pacific Grove. The apotheosis of the questing friendship between the three men and Steinbeck's wife, Carol, came by way of Jeffers. Bursting into her house she exclaimed to the three men: "I've got the message of 'Roan Stallion'!" Each of them recognized in his own way that Carol had put her finger on exactly the problem of mankind and exactly its remedy as articulated in Jeffers' poem:

> *Humanity is*
> *the start of the race; I say*
> *Humanity is the mold to break away from, the crust to break*
> *through, the coal to break into fire*
> *The atom to be split.*

Robinson Jeffers is the poet of "inhumanism," arguing that people need to uncenter themselves from the picture of life on earth and understand themselves as part of a teeming multitude of other living things, part of a natural world that existed before *Homo sapiens* and will exist after us. Jeffers' work is a focal point for the Dark Mountain Project, an amalgam of artists, writers, and activists in England who state that they have "stopped believing the stories our civilization tells itself. We see that the world is entering an age of ecological collapse… [and] that writing and art have a crucial role to play in coming to terms with this reality."[37]

In "Roan Stallion," Jeffers echoes Herman Melville but subverts Ahab's dark directive about breaking through the mask of the visible world to kill the life force behind it. Jeffers finds affirmation where Melville locates destruction. It was Jeffers' version of the mask and not Melville's that would inform Campbell's *The Mask of God*: "Slit eyes in the mask; wild loves that leap over the walls of nature." There is something in Jeffers' poem for Steinbeck, too, who constituted the phalanx from, in his own words, "arrangements of atoms plus a mysterious principle."[38] The stallion itself, merged with mythic dimensions, is an explosion of beautiful life out of matter, and tragically sacrificed to human ego assertion, which is "the mold to break away from." Ricketts took the title of his essay "Breaking Through" from the poem. Campbell quoted it his whole life.

The Circle Is Broken

When later in life Campbell talked about this time, invariably he painted it in ecstatic terms. "We were all in heaven," he said. "The world had dropped out. We weren't the dropouts; the world was the dropout. We were in a halcyon situation, no movement. Just floating. Just great . . . So I'm coasting along, trying to find where I am. . . ." Campbell recalled experiencing kairos, the merging of the single moment in time with the all of it—the supreme moment. In a separate instance of explaining his own kind of flow experience, Campbell put it this way: "It was like a symphony, with everything coming from a single center that I was hearing and discovering."[39]

The happy group was soon to break apart mainly due to what was perhaps a full-on affair or perhaps just a sweaty flirtation between Campbell and Carol Steinbeck. To the reader going along with Campbell's biography, this dalliance comes as something of a puzzle. Campbell was by many accounts fabulously good-looking, and women did toss themselves at him with regularity. He would seem to have mostly resisted. He managed long friendships with women who at certain points in their acquaintance would likely have been quite pleased with more from him.

He was given to scrupulous behavior, not a casual betrayer. "We were all more or less in love with each other, I guess," Campbell wrote in his journal. "It was almost like a little fugue of loves."[40] There was a public scene in which Campbell and Carol made out rather at length outside a party, of course attended by everyone else in the gang, including Steinbeck, who watched for a moment then returned to the house "in disgust."

Long discussions and resolutions ensued, including several civil if fraught confrontations between Steinbeck and Campbell. These people were all young—Ricketts was the senior statesman at thirty-five. They were thrashing around together and figuring things out. Campbell had found himself the "experience" at the heart of art and of the hero's journey. Following the medieval script, now he would play Lancelot to Steinbeck's Arthur and declare his exalted love.

Campbell confessed to Steinbeck that his love for Carol was beyond physical desire. "Maybe you are in love with an idea that isn't Carol at all," Steinbeck replied. And so they talked, and drank coffee. According to Shillinglaw, Steinbeck breached the impasse, in June 1932 ordering Campbell "out of town at the point of a gun—literally."[41]

Ricketts was in his own hot water at the time, and was attempting to disentangle himself from an ill-advised affair with the seventeen-year-old Xenia Kashevaroff, sister of Tal Lovejoy and Sasha Calvin.[42] (Further cultural currents burbling here—Xenia became a nude model for the photographer Edward Weston and was long married to, and then divorced from, the musician John Cage. Cage wound up with Merce Cunningham.) Ricketts was about to embark on a collecting trip to Alaska with Jack and Sasha Calvin aboard the thirty-three-foot *Grampus*, a former navy launch, and the idea came up for Campbell to go along. After all these theatrics, it was time to address the reality of the tide pools again. And get out of town.

Decades later Campbell reminisced about the trip. "We would go out in a little canoe that Jack had, and every time there'd be a good low tide place to gather them, we were gathering *Gonionemus*. It was a beautiful trip.... Nobody had any money.... For the four of us it was twenty-five cents a day on that boat, including the gasoline."[43] When they arrived in Sitka, "the dock stank like fish and they could see the green cupola of St. Michael's Russian Orthodox Cathedral from the harbor." They spent almost three weeks "canoeing, reading, listening to Stravinsky, snapping photographs and taking long walks" through a park of Northwest Coast Native totem poles.[44] Ricketts typed up his collecting reports and Campbell read Dostoyevsky. They talked into the night. Calvin and Ricketts worked on *Between Pacific Tides*, and Campbell intermittently proofread it, "chiefly correcting the grammar," which did not endear him to Calvin.[45] In these ten weeks, Campbell's nascent ideas were more deeply imbued with biological dimensions. And while the hero's journey has plenty of other historical influences, in its outlines can be discerned Ed Ricketts' major preoccupation with his concept of "breaking through."

Ricketts was never terribly clear about what that meant, and in a sense, Campbell defined it for him. In Campbell's interpretation, biologically founded myths are the "masks of God" through which men everywhere have sought to relate themselves to the wonders of existence. So Campbell might have explained to me as a child that although eating the host to experience Jesus is, yes, a myth, that's its strength, not a weakness. The ritual is an entry point to transcendence, from which opening is revealed the accessibility of the supernatural through every object everywhere.

"The mystery of life and consciousness pours in through the various bodies and beings round about. It must then show you yourself that you are similarly transparent to transcendence."[46]

The connections between biology and spirituality are right there in the tide pool: "All those strange forms," Campbell said, "cormorants and little worms of different kinds and all. You'd hear, my gosh, this generation of life was a battle going on, life consuming life, everything learning how to eat the other one, the whole mystery, and then from there they crawl up on the land. And also in the mythic themes generally out of the ocean, or what in India is called the milky ocean out of which the whole universe comes."[47] Campbell described myth naturally issuing forth: "the full mystery of the teeming life of the earth is contained within the egg of a flea," a "production of the living psyche."[48]

Regeneration

The Great Depression was a time for young men to drift and dream, and for their fathers to lose businesses. Campbell's West Coast adventure was curtailed by a telegram announcing this fate had befallen his family back East. Campbell had contributed greatly to the creative ferment of Pacific Grove. Shillinglaw calls the time he intersected with Steinbeck and Ricketts "the generative time, perhaps the *annus mirabilis*" for all of them.[49] But Campbell had also disrupted the ecosystem. Upon his departure, relationships recalibrated and proceeded apace. Campbell would work on fiction writing for years before gradually becoming the sage and seer of world mythology.

In contrast, Steinbeck was on a long roll, producing a novel a year—a partial list includes *Tortilla Flat* in 1935 and *In Dubious Battle* in 1936. In a five-month creative marathon beginning in May 1938, he birthed *The Grapes of Wrath* and collapsed at

Ricketts' door basically upon completion. This novel vies with 1952's *East of Eden* as his greatest accomplishment. From the perspective of land use, climate change, and the social inequities of our great democracy, the creative insights of *The Grapes of Wrath* continue to resonate. This big novel neatly encapsulates the consequences of human dissociation from the land, leading to destruction of that land, and then to the destruction of humanity itself—a theme that is playing out, big time, today.

The Grapes of Wrath changed Steinbeck's life. He was roundly praised and roundly reviled for the work; the whole thing exhausted him. Those who loved the book called his vision of social justice revolutionary; those who hated it focused on the same theme and called it cardboard. He was accused of communism and stalked by the FBI. It was a huge best seller yet distasteful to perhaps the most influential critic then opining in the United States. Writing in the *New Republic*, Edmund Wilson called out Steinbeck's characterization of man in animal terms. "This animalizing tendency of Mr. Steinbeck's is, I believe, at the bottom of his relative unsuccess at representing human beings."[50] Wilson would perhaps revise his opinion today given what we know now about the realities of the Anthropocene. *Homo sapiens* has entered a new phase of our reality as "group man," and many people would say humans make other animals look downright civilized by comparison.

Plenty of other pressures had piled up on both Steinbeck and Ricketts, making the idea of an expedition to a beautiful place very attractive indeed. Steinbeck's wife Carol had held his hand through the intense process of writing *The Grapes of Wrath*. She typed the manuscript—no small feat—and came up with the title, but things between them were deteriorating. Ricketts' marriage, such as it was, had reached a definitive impasse and he was once again prepared to flee romantic fallout. As much as it was a voyage of discovery, the trip was also meant to heal hearts and minds. And to get out of Dodge.

Steinbeck declared he was done with fiction and that the "new thinking" was the study of science. He accumulated pages for a planned but never-finished marine guide to San Francisco Bay under the guidance of Ricketts. This ambition was replaced with the idea of a collecting expedition around the Gulf of California (also known as the Sea of Cortez). The unique and rich waters surrounding the Baja California Peninsula had not yet received major scientific attention, and the specimens Steinbeck and Ricketts collected remain important vouchers in use by scientists to this day.[51] Half their specimens were new to science, not yet named or described, or were collected significantly outside their known ranges. At Steinbeck's suggestion, the trip was conceived as a literary as well as a scientific enterprise. Putting a proposition to Ricketts, he said, "We'll do a book about it that'll more than pay the expenses of the trip."[52]

The physical voyage officially commenced on March 11, 1940, with John and Carol Steinbeck, Ricketts, and a four-man crew aboard the *Western Flyer*, a seventy-six-foot sardine purse seiner. A party atmosphere attended the launch and the *Monterey Peninsula Herald* reported on it. From Monterey they headed to San Diego, then made their way down the peninsula to Cabo San Lucas at the southern tip (arriving at Cabo on March 17). The *Western Flyer* then proceeded up the peninsula, inside the gulf, steaming as far north as Bahía de Los Ángeles before crossing to the east side of the gulf, and then south for the return to Cabo San Lucas. The expedition completed its tracery on April 18, docking once again in San Diego.

Sea of Cortez: A Leisurely Journal of Travel and Research was published in 1941. Comprising a narrative of the journey and a species list, it ran to six hundred pages. Steinbeck's editor, Pat Covici, suggested the work be attributed thusly: "By John Steinbeck. With a scientific appendix containing materials for a source-book on the marine animals of the Panamic Faunal Province, by Edward F. Ricketts."[53] This did not go over. Steinbeck hotly insisted that the book was wholly the product of both men, and it was copyrighted to reflect dual authorship. Steinbeck had done the actual writing of the book, based mostly on a journal kept by Ricketts; as the authors clarified for Covici, "In one case a large section was lifted verbatim from other unpublished work." What has become known as the "Easter Sunday Sermon" in *Sea of Cortez* is Ricketts' essay on non-teleological thinking, the conclusion of which states the intention of book and journey: "This little trip of ours was becoming a thing and a dual thing."[54]

The Toto Picture

From the logs of Gaspar de Portolá to those of Captain James Cook and Lewis and Clark, the expedition journal has become an important category bridging history and literature. These accounts inventory the past when it was a present. They are the testimony of the eyewitness. Studded like precious gems in the narratives are corrective visions of what constitutes a place that inform how we see that place today. While Alexander von Humboldt and Charles Darwin shaped their travel narratives to literary ends, they did not venture onto their journeys for expressly artistic purposes, which Steinbeck and Ricketts in part did. In going on the trip and in writing the book, they sought to fuse existing forms into something new, a narrative that did not just document observations from A to B and back again, but provide an experience in itself, a "breaking through."

To write this book, Steinbeck declares in the introduction, "we have decided to let it form itself: its boundaries a boat and a sea; its duration a six weeks' charter time; its subject everything we could see and think and even imagine." Their avowed purpose was to locate, observe, count, and collect marine invertebrates, "and when we used this reason, we called the trip an expedition." This was only part of the truth, he admits, because the main reason they undertook the trip was that they were curious. Steinbeck's narrative mode is to repeatedly reach out for the grand, then to reveal that what he actually has in his hand is a modest thing. In this way *Sea of Cortez* is paradoxically something of an anti-expedition narrative, since its authors never take a dominating or appropriating stance toward the territory they traverse, but rather integrate observations from an essentially receptive position.

Invoking among others Linnaeus and Darwin, Steinbeck said that as they did, he would build a structure "in modeled imitation of the observed reality." The *Systema Naturae* and the theory of evolution by way of natural selection are rather daunting models. Steinbeck and Ricketts philosophize, opine, and reflect, and as promised, repeatedly return their attention to what Ricketts called "the good, kind sane little animals."[55] As they strive after pattern recognition, the going out of thoughts and the pulling in of direct observation convey a tidal rhythm in the text. The "moon-pull" is implicitly mimicked and explicitly assigned to all manner of phenomena. Steinbeck wonders if the same tidal force that incites the gametes of polychaete worms to explode may mysteriously cause the Old Man and the Sea to rise before fishermen "in the pathways of their boats." The vision may not be a hallucination but an experience joining "past and present" through a rhythm that pulses in both the physiology of creatures and the symbols of the mind. Near the end of the text, Steinbeck concludes, "The laws of thought seemed really one with the laws of things."[56]

Joseph Campbell understood immediately what they were trying to do, writing to Ricketts, "I think that the book form discovered by you and John is perhaps as close to the life-form itself as [a] book could possibly be to life."[57] He told Ricketts, "During the past six months the world of the myths has been revealing to me those simple, wonderful forms that underlie and sustain the bewildering pell-mell." He promised to send some papers on the subject in a few years, "written in the jargon of another science."[58] The result almost a decade later was *The Hero with a Thousand Faces*. It would provide another structure for discerning the pattern of "observed reality." And in some ways it would more fully articulate the yearned-after sense-making of *Sea of Cortez*.

In 1951, at Pat Covici's urging, the book was reissued without the species catalog and retitled *The Log from the Sea of Cortez*, by John Steinbeck. Things had changed in the ten years since the book was first published. Steinbeck had moved to New York. Early in the evening of May 8, 1948, Ed Ricketts had headed out of his lab to get dinner fixings and, crossing railroad tracks in his car, was blindsided by the Del Monte Express. He died three days later. Steinbeck chronicled his grief and his friendship in "About Ed Ricketts," an essay that is presented as an appendix to the *Log*. Various Ed Heads who carried and still carry the Ricketts flame look upon this essay as a betrayal, finding "About Ed Ricketts" an inadequate portrait of the man and the severed *Log* a suspicious maneuver to give the best-selling Steinbeck full authorship and so boost sales of the book. Others admit that the story is complicated. Shillinglaw pointed out to me that the species catalog was unwieldy, after all, and that Ricketts always insisted Steinbeck had done nearly all the writing. Steinbeck "thought that with Ricketts's death, it was a different project," and had become explicitly about their friendship.

Three Cases of the Ricketts

" *You* are one of those Ed Ricketts nuts?" I asked Richard Brusca, who just doesn't fit the funky mold of the tribe at all. Unless you count besotted devotion to his wife, Wendy Moore, who is a professor of entomology at the University of Arizona, Brusca is very un-nut-like. "You are a drill-down evolutionary biologist in the traditional mode, Rick!" He laughed. Guilty as charged. Now executive director emeritus of the Arizona-Sonora Desert Museum, since his official retirement Brusca has been putting in more than full days at his desk. He has been turning out, among other things, peer-reviewed papers, including one on plant species moving in response to climate change; a natural history guide to the sky island region, mostly in Arizona (Moore coauthored); and most recently a third edition of his best-selling college textbook *Invertebrates*, which has been in print since 1990 (in four languages). I have stayed at the Brusca-Moore household and witnessed him at work—you can practically discern phyla, genera, and species thought balloons above his head.

"I grew up with marine biology," Brusca told me. "I taught with Joel Hedgpeth and we became lasting friends." Hedgpeth being the iconoclastic but also traditionally academic marine biologist who edited several editions of Ricketts' *Between Pacific Tides* as well as more Ricketts prose in *The Outer Shores*. (Hedgpeth was among those who felt Steinbeck did not give Ricketts his scientific due.) "I grew up imbued with a sense of Ricketts. I was embroiled with people who had known him and were still working—they were in their sixties. We all just referred to him as Ed." Listening to Brusca, I got the sense, as Faulkner surmised, that the past isn't even past. "Ricketts was at that interface between citizen scientist and real scientist—and by the time he died he was so respected. He codified the now universally accepted concept of intertidal zonation and was far ahead of his time."

Brusca is among those who have retraced the 1940 route of the Western Flyer. In 1972, Brusca and J. Laurens Barnard of the Smithsonian Institution visited the twenty-three field sites where Steinbeck and Ricketts collected, over a period of seven weeks (Steinbeck and Ricketts took six weeks). Eventually Brusca paid Ricketts forward, producing what Hedgpeth praised as the "monumental 1980 vade mecum, *Common Intertidal Invertebrates of the Gulf of California*."[59] Brusca told me that Ricketts' and Steinbeck's collections were "not quantitative, empirical, or replicable in a scientific sense. Ed and friends went out when the tide was low." This means they did not geolocate areas of inquiry or develop a line of thinking along which they collected. "It was an expedition of discovery at the time, which was needed, but nothing was done in a quantitative way." Brusca has completely cataloged all their specimens from the Sea of Cortez expedition and these number about five hundred species. As a second-generation Ed Head and, yes, drill-down evolutionary biologist, Brusca brought Ricketts' work into a more organized and quantitative form.

A Serious Lark

Nailing down the who, the what, and the where is the directive of the scientific inventory but of course Steinbeck and Ricketts were up to something more. Stanford University biology professor William Gilly has long taught a course in "holistic biology," defined broadly along the lines of the "toto picture" aspired to by his gurus. "We use marine ecology as a vehicle for exploring complicated problems in the real world, like anthropogenic change, social human interactions—how all systems are coupled," Gilly told me. "It is based very much on *The Sea of Cortez*—that's our bible. It includes a lot of important philosophy about how things work, and something that is obvious but not taught very much to students: the natural world includes people."

In 2004 Gilly organized an expedition expressly to follow the example of Ricketts and Steinbeck. He teamed up with Jon Christensen, today a professor at UCLA and the editor of *Boom: A Journal of California*. Both a journalist and historian, Christensen was well equipped to amplify the trip's literary dimensions and told me it was "a serious lark."

In a kind of "everybody onto the boat" spirit, Gilly invited Katharine Rodger to join the expedition for a part of it and she brought along her mentor, Susan Shillinglaw. Falling in love on the journey, today Shillinglaw and Gilly are married, and Shillinglaw co-teaches the holistic biology class at Stanford. Gilly, Christensen, and a core of other seekers set out to methodically revisit fifteen of the original field sites, and they added fifteen others. "After studying these thirty sites by laying out transects and counting every visible organism in regularly spaced half-meter square quadrats, I am mostly tired but still curious to decipher the changes," writes Gilly in a journal-style article after the expedition. Eventually Gilly would inventory more than six hundred quadrats with the intention of setting a baseline for future studies. "So much depends on your point of view and a commitment to keep looking." The physical changes are pretty much uniformly dispiriting and include a drastic reduction in the abundance of most species, and a huge increase in development, as in Cabo San Lucas, which Gilly calls "the adult Disneyland of Baja." He adds, "No habitat means no animals to count." The richest location in 1940, a harbor at Puerto Escondido, is today "half concrete."[60]

Documenting evident range changes in giant squid, far more abundant now than observed by Steinbeck and Ricketts, Gilly addresses the seeker's discovery—which is not necessarily to repeat precise observations but to see anew for oneself. "To be alone in search of squid, and to have found it all, seemingly in the real Rickettsian holistic sense, has made some of us feel so much closer to the original spirit of the 1940 trip," he writes. "At this point we are just following our own course and seeing with fresh, new eyes. This vision that comes with escape is why I wanted to undertake the whole trip in the first place and was all I ever really hoped for."[61]

Gilly has gone back and continues to go back. He's involved in citizen science efforts to monitor the Santa Rosalía coast of Baja, which he says is "spectacular and undeveloped—like it was in the 1940s." This is part of Mexico's biggest protected area, but unless there are baseline inventories of what lives there, it will be impossible to keep protecting them. He has also helped set up intertidal transects and surveys for monitoring the impacts of a gigantic mining operation in Santa Rosalía. Here, if you build it, invertebrates will come, and have already availed themselves of new habitat created by boulders mined from the mountains and dumped in the water. A pier under construction had so many lobsters congregating around it the locals started to sneak in to fish them. Discussions are underway to simply create another artificial reef outside the mining area so that the locals don't have to trespass to get to the lobsters.

Wish You Were Here

One sojourner along for some of the Gilly-Christensen voyage was Rafe Sagarin, a young biologist mentored by Brusca and others. Sagarin personified the third generation of Ed Head scientists along on the expedition. Not so many years later, Sagarin was hit and killed by a drunk driver while riding his bike in Tucson, Arizona, in early 2015. He was forty-three.

"We were in a café near the Burma border in Thailand, where

we were collecting insects, and Wendy was checking her email," Brusca told me. "She told me I'd better sit down, she had bad news. It took me several days to absorb it. I still haven't really absorbed it." Brusca described Sagarin to me as an effervescent, outside-the-box thinker and dreamer. "Some might call him a dilettante, but he didn't allow himself to be constrained by the way university politics work today. You have to have a big body of research in a very specific topical area so that you are a world expert in that field—but he was just too broad for that." His job at the University of Arizona's Biosphere 2 was not your traditional academic gig.

Sagarin was the author, with Aníbal Pauchard, of *Observation and Ecology: Broadening the Scope of Science to Understand a Complex World*.[62] "This book," the authors declare, "is dedicated to the goal of recovering a respect for excellent observations of nature." Sagarin and Pauchard discuss the contributions and benefits conferred by citizen science, and they are strongly in favor of it. As they put it: "The simplicity of going out into nature and counting, measuring, watching, and recording opens ecology to a non-elite, nonprofessional world where people who don't spend their lives as ecologists can nonetheless contribute to ecological science" and can then disseminate their ideas "into other endeavors, like politics and art."

Sagarin and Pauchard argue that the hypothesis-driven, experimental methods of most peer-reviewed science have serious limitations, especially when dealing with ecosystems, which are complex, confusing, and involve nonlinear interactions across multiple scales. Their book restlessly challenges academic science for plucking observations of nature out of context and analyzing them with technology and according to abstract models. Hypothesis-driven science in general relies on what biophysicist John R. Platt termed *strong inference*, or a yes/

no, reject/do-not-reject testing of one hypothesis against another until the most plausible one stands. Sagarin and Pauchard contend that this method "relies on testing a continually dividing tree of binary hypotheses, but many ecological phenomena occur across a continuum." To grapple with the multiple scales of time and space at play in nature requires first and foremost "a straightforward call to simply look at the data." Their book advocates for a more "toto picture" approach based on observation.

At his death, Sagarin had a paper near publication, which Brusca is a coauthor on, about "mesocosms."[63] As Brusca explained it, "In the experimental world, there are three levels. The tightly controlled chamber in your lab is one. Another is the real world." Brusca's real-world example is his recent botany paper on the Santa Catalina Mountains in Arizona, in which he shows plant species moving significantly upslope in recent decades, tracked right alongside climate change data. Brusca was able to do this analysis because sixty years ago, seminal ecologist Robert Whittaker inventoried the same mountain range, and Brusca could compare what he found with what Whittaker found. Whittaker's documentation was not as precise as would be required today, but Brusca could still geolocate his observations with general accuracy and go to the same spots. Historic data sets like Whittaker's are in very short supply, and creating them to capture life as we find it today, and constantly updating them to track changes, is a main imperative of citizen science.

"But the third scale," Brusca explained, "is between the microcosm in the lab and the macrocosm in the world, and this was Rafe's idea. To create a mesocosm of the Gulf of California right here at Biosphere 2." Biosphere 2 is a domed construction outside Tucson that is now owned and used for scientific research by the University of Arizona. It was originally constructed between 1987 and 1991 as a self-enclosed system replicating earth's biomes; the experiment was basically to see if such a thing could be done. In sum, it could not, at least at the time. Based on what has been learned from the original experiment, the idea remains feasible. And some nifty research occurs on site. "Rafe's idea was to turn the ocean element mesocosm into the Sea of Cortez." This is an ambition of which no doubt Steinbeck and Ricketts would approve: the mesocosm benefits from the focus of a lab yet allows for the large-scale dynamism of nature unfolding on its own terms. Though at this point in his projected plans he hadn't articulated how he envisioned it to work, Sagarin's ambition for the Sea of Cortez mesocosm included citizen science or, in his terminology, "public participation in science."

Rafe Sagarin went more than once to the Gulf of California in the spirit of Steinbeck and Ricketts. Writing about a trip with students conducted a few years earlier, he noted, "I wanted us to go 'doubly open,'"[64] as Steinbeck and Ricketts declared they were doing. Sagarin's "doubly open" entailed "a whole spectrum of observation between the coldly scientific and the deeply experiential poles that Steinbeck and Ricketts staked out." He wrote, "I've come to see science as a shifting, evolving thing," not calibrated upon "astounding discoveries" but rather occurring "holistically in system-wide modifications of simple, ancient processes...driven by observation."

The Art of Not Science

"Why are you calling Ed Ricketts a 'citizen scientist'?" is something I have been asked frequently by myriad filmmakers, writers, and other artsy types who are among Ricketts' followers today. "He was a real scientist. It was his job." Actually, "real" scientists, even when they are Ricketts fans, don't give him the title. "Not a scientist," Pete Raimondi told me about Ricketts, even while admitting he pursued his own vocation as a marine biologist partly because of Ricketts' example. "He was a collector." Even Ricketts' conceptual organization of the tide pool, credited though it is as an important ecological contribution, blurs the line between observational and quantifiable knowledge. Tide pool zonation, the sorting of different creatures into their various niches, has overlaps, exceptions, complexities.

Why is Ricketts so important to you all, then, I asked Rick Brusca, if he didn't do real science? Brusca quoted me oft-repeated Steinbeck: "It is advisable to look from the tide pool to the stars and then back to the tide pool again." Then he explained, "Sometimes scientists have the urge to connect to the universe. When you read Steinbeck and Ricketts, that's what they talked about all the time. They have become gurus in some context we unconsciously relate to and follow in some way—which is why the quote is repeated so often. It encapsulates the wanderlust most scientists have, but we get wrapped up in the minutiae and the statistics, and we forget it. Then you are sitting by a campfire and looking up at the stars and you feel it!"

Brusca told me that one of his most significant scientific contributions came about by way of a more observational approach à la Ricketts. His example also explains, if you have been scratching your head about it, how a marine biologist came to be boss of a museum in the desert. Part of the answer is that he loves both environments, and another part of the answer is that the two go together. "It dawned on me that the Sonoran Desert is the way it is because of the Sea of Cortez—all our summer monsoon rain is from the gulf, and that's what gives us our incredible diversity here. The big picture is that we get a huge transfer of heat and water from the gulf—down to the carbon made by seabirds from the gulf that leave feces and corpses here." Brusca figured this out in an intuitive way; he said it "dawned on" him.

He coined the now commonly used term *maritime desert*. "If you are a real observer," Brusca said, "you look at more than one thing at a time and at different scales at a time. But if you want to get a paper published, you'd better translate all that into analyzed statistics."

Sagarin put it similarly, reflecting that at the end of the trip with students, "I got to thinking about the long chains of observations, and scientific knowledge, and the links that are forged" between students and teachers and between historical figures and those who read and follow the work they left behind. Again by the campfire, both real and proverbial, Sagarin said, "we . . . listened to the waves crashing on the rocks and watched the flickering embers rising to meet the flickering stars, and I thought how everyone's favorite moment had become my favorite moment, too. So in that moment I was finally taken to that place that existed only on paper when we started the trip, a place mapped out in prose by Ed Ricketts and John Steinbeck on one of their own leisurely journeys of travel and research." Sagarin was breaking through to Ricketts, and vice versa. And having put a distinctive marker of his own on what it means to journey to the Sea of Cortez, perhaps Sagarin can be invoked as Ricketts is when subsequent seekers follow. I think "citizen scientist" is a compliment Ricketts would gladly accept. *Scientist* in general refers to a man or woman alone, and *citizen* is communal—not only as one member among many in one place and time, but across those boundaries as well.

CHAPTER TEN

Bee, I'm Expecting You

There is a guru of citizen science, and it is Sam Droege. The bioblitz–a rapid inventory of species undertaken by the citizenry usually in a national or state park–is just one of the products of his prolific and generous imagination. Sam Droege also happens to be someone whom you can ask, "What in *h-e-*double-toothpicks is going on here on this benighted earth and what are we going to do about it?" and you will get a fascinating and maybe kinda sorta consoling answer.

Journalist Andy Isaacson likens Droege to Johnny Appleseed, a comparison that, given the mythic figure's association with the starring fruit of Eden and natural history's theme of a dispersed paradise, is apt.[1] One quibble, however. Appleseed was bent on cultivation–he planted orchards–and Droege's attentions are trained on the wild. Droege is a serial inventor, and he's giving ideas away all the time. "I feel like it's my role to say, here's something we developed, here's how it works best, now you run with it. We've done a whole bunch of things that don't work too so I let people know about that too." Among the projects Droege has instigated are the North American Amphibian Monitoring Program, FrogWatch USA, and Cricket Crawl. Currently he's immersed in an effort to inventory wild bee species in the national parks.

This kind of sounds like a list of darling-species watches in which earnest nature nuts try like mad to shore up disappearing traces of biodiversity like kindly elves helping out at a wake. While the specificities of the life stories and progress or lack of it for birds, bugs, frogs, and so on are of the utmost relevance to figuring out what's going wrong in nature, a bigger view comprising all the minutiae is necessary for really grappling with the issues we are faced with, and Droege knows this full well.

In February 2015, Droege spoke at the Natural History Museum of Los Angeles County. The house was packed; beyond the doors of the lecture hall, thousands of young people mingled and danced among the exhibits at the museum's First Friday evening event. The museum has a panoply of citizen science projects going, most of which are connected to the research agendas of scientists working at the museum. The theme for the 2015 First Friday lectures was "Do-It-Yourself Science"; the series description said, yes, there are scientists who "make a career out of asking big questions," but that there are also great scientists, from Benjamin Franklin to Isaac Newton and of course our darling Darwin, who "approached science not as a job but a hobby," and this journey of discovery can be taken by any of us. "With the rise of the 'citizens' and the 'scientists' working collaboratively, we can revolutionize and redefine scientific exploration."

Droege took the stage in a red shirt sparkling with rhinestones, which set off his white ponytail and beard. (Today Droege is beardless and sports Viking braids.) One of his main points to the audience was that those vaunted big names are not actually so important. "If they hadn't published their findings, someone else would have. I'd argue that if Thomas Edison hadn't existed, we would still have lights on here right now." Their significant gifts to society would have been conferred by someone else and added to and supplanted by others. Droege said the real useful addition to science is made by the mostly anonymous citizens who contribute observations and data points "in a time and a place that can't be replaced." He conceded that Joseph Grinnell, who established guidelines for taking really useful field notes that make it possible for researchers to resurvey his data and so figure out situations like climate change, is Someone Who Made a Permanent Contribution.

The next day a friend and I picked Droege up at an unassuming hotel in Hollywood; he had requested that the museum put him up in this location because from here he could walk to Griffith Park and do some bee hunting. Ergo, Droege appeared with a gigantic net, and off we went.

"People see the net and they know immediately you are not a threat," he said. We traipsed around Griffith Park and talked about the mountain lion who has made his home here, getting across eight lanes of interstate to do so—talk about the hero's journey! The story of this mountain lion, called by prosaic biologists "P22," has inspired the National Wildlife Federation to begin planning a wildlife corridor connecting Griffith Park to the Santa Monica Mountains, so that others in his tribe can make the journey to and fro without getting hit by traffic. It will be the largest wildlife corridor in the United States. As positive as this outcome may be, the brave P22 will not get to complete his biological destiny; while he has lots of deer and small mammals at his gustatory disposal, he will likely die single and without offspring. You can follow him on Facebook, "P22 Mountain Lion of Hollywood."

Only Connect

The science of wildlife corridors has its origin in island biogeographic theory and was given a significant boost by William Newmark in the late 1980s. Newmark, today a research curator at the Natural History Museum of Utah in Salt Lake City, was a young man keenly tracking ideas about how to design reserves to stem extinction. Controversy centered on "the question of whether or not island biogeographic theory really applies to nature reserves," Newmark explained. After all, nature reserves are not literal islands surrounded by water.

Working in Yellowstone and Grand Teton National Park on his master's thesis, Newmark considered that since national parks are discrete nature reserves, like islands, and have species records documenting who lives in them, "you could actually test directly whether IBT applies to extinction in historical versus geological time." Choosing twenty-four national parks similar in age, biological region, and historic mammal communities, in 1983 Newmark hopped in his old Toyota station wagon and set out to visit them. "I slept in it," he laughed. "It was my RV." Once at the parks, Newmark pored over sighting records made by biologists and park employees. He put the marker for local extinction of a species in a park if it hadn't been seen there for at least ten years. "I thought it was conservative to say that if you haven't seen a species in the last decade," it's locally extinct, he said.

Once he developed lists of who was still where and who was gone from each park, Newmark could look at patterns of extinction. "Smaller parks should experience higher rates of extinction than larger ones. Older parks should experience higher numbers of extinctions than younger ones, and the number of extinctions should exceed colonization rates in all parks. Lo and behold there was support for all three predictions."

Newmark showed that forty-two populations of species had gone extinct in the parks he surveyed, and that "virtually all western North American national parks were too small to maintain the mammalian faunal assemblage found at the time of park establishment."[2] In the paper Newmark suggested the parks be augmented by adjacent lands. Today, the "connectivity" concept is perhaps the single most powerful takeaway from the lessons taught by IBT. Fragmented landscapes, even when the fragments are huge, like some of our national parks, lead to extinction. Connecting wild landscapes gives far-roaming species a way to fulfill their territorial imperatives and their need to disperse; if limited spatially, populations experience genetic inbreeding depression and blink out.

One-Off

In Griffith Park, Droege hunted for native bees. We found none, because we were talking too much. Several people stopped us and asked about the net. One concerned mother told Droege she has her kids eat bee pollen to bolster their immune systems. He advised against this. "All the pesticides and junk they're putting out there are concentrating in the pollen," he said.

Droege started out as many naturalists do, crazy about birds, and under the tutelage of his longtime mentor, Chandler Robbins, he took charge of the Breeding Bird Survey in 1986.[3] Chan Robbins is still at it with the birds at age ninety-seven, his vintage evoking "the era of the great naturalists... a time when our nation's woods, fields, streams, and skies teemed with unexamined wildlife," according to a profile in *Audubon* magazine.[4] In the middle of the environmental paroxysm of the early 1960s, Robbins, a US Fish and Wildlife Service biologist, received a letter from a birder, asking a simple question: Was the continental robin population in decline? Robbins considered that without a scientifically rigorous, continental monitoring program there was no way to know. He conceived of a roadside point system that would be easy for car-loving volunteers to follow, randomly selecting routes based on USGS topographical maps. He partitioned states into blocks of one degree of latitude and longitude and then randomly selected uniform numbers of starting points for each block, also determining the direction of the route randomly. Left to their own devices, birders would tend to go where there were usually the most birds to see, likely passing over the places where change was occurring—less attractive habitats being disturbed by some impact or other. Robbins' standardized methodology and sampling protocols were what yielded statistically useful data.

This was all manna for Droege, who is a statistics nut. He is also a poetry nut. I'm on two of Droege's listservs, and every morning I try to open his daily poem email before I let the world of "anger and telegrams" sully my brainpan. (E. M. Forster would no doubt reference "anger and emails" today.) How much better to set one's course by Pablo Neruda. From *Art of Birds*:

> I've wandered the world in search of life:
> bird by bird I've come to know the earth
> ....
> even though no one paid me for it…

I am also one of nine-hundred-plus members of Droege's bee-monitoring group. As evidenced by his overall gentle-conservation-biologist demeanor, Droege is a peaceful keeper of bee people, but sometimes he has to get firm. One snarky and personal exchange tipped off by a tussle over the relative stupidity of using bowl traps to catch bees escalated to the point where Droege expressed his opinion that the public conversation was not advancing science and so ought to desist. His flourish came by way of Emily Dickinson: "A Word dropped careless on a Page / . . . Infection in the sentence breeds."

After six years with USFWS, Droege moved on to the USGS. Based at the Patuxent Wildlife Research Center in Maryland, here he resurrected what is now called the North American Bird Phenology Program. In 1880 a schoolteacher with a serious thing for ornithology, Wells W. Cooke, solicited observations from other birders about what species were arriving when and leaving when in their neighborhoods. These observations were handwritten, of course, on cards, and mailed to Cooke. Some of the cards in this collection include contributions by legendary naturalists like Joseph Grinnell, Aldo Leopold, and Margaret Morse Nice. Theodore Roosevelt sent in observations from the White House lawn.

"Chan and I had, at various times, kept the cards from literally being put in the Dumpster" by bureaucrats, Droege told me. "We rebranded the program and it became one of the first groups to crowdsource data entry for science online. We might have been the very first to do that, I'm not sure."

The Bird Phenology Program was discontinued in 1970 and nobody quite knew what to do with the cards. Today they are being slowly digitized under the diligent care of researchers, first Jessica Zelt and now Eric Ross. "We can't hire people to do it but we put up images of these cards" online, Droege explained. Citizen transcribers copy the data from the cards and each card gets transcribed twice. Droege explained that the job is tricky because misplacing a comma or omitting an extra space can produce a mismatch. Once the card is verified, its contents become useful data points. Several people have transcribed more than one hundred thousand records each. The project is far from done and you can help: www.pwrc.usgs.gov/bpp.

Droege has some main takeaways for how to do effective citizen science. He said one reason people keep transcribing cards for the North American Bird Phenology Program is the personal connection with its leader, Chan Robbins. Personally, I find this piece of insight completely right, and I would actually add that a charismatic leader, beyond even a more generally likable one, is a critical piece of what keeps a citizen science project going. The most robust citizen science project I participate in, the Golden Gate Raptor Observatory's Hawkwatch, has more than three hundred volunteers year after year counting and banding hawks not just because of the birds but because of the program's leader, Allen Fish. Droege also counseled that no one should ever ask permission to start a monitoring program: just do it.

Though he is an obvious booster for it, Droege cautions against counting on the masses to really get nature's jobs done. "There's a lot of interest in citizen science projects when they start up," he told me. "People attend a meeting. They do something once or twice. The bulk of what gets done is by a small set of fanatics." Droege told me about Jane Whittaker, a volunteer on his bee survey who has more or less single-handedly inventoried the native bee population of West Virginia. "She's a force of nature," Droege said. Availing herself of the image she projects of a harmless older lady, "she walks into farmers' fields and chats them up. Somebody else would get arrested for trespassing. She washes, dries, and pins the bees at her house, and does a preliminary ID. I double check it and do the data entry here. Every year we set out a plan based on her capacity and interest. She's collected more than twenty records of 'never before seen in West Virginia.'"

"Find the core group of people who are really going to do the work, and cultivate them," Droege advised. "It's key for people to feel they are part of a group even if they never see anybody else. I email my volunteers, send them poetry, little interesting facts and pictures, and a million reminders. How do we recruit a sense of community across boundaries when we no longer live in villages? Establish an internal sense of obligation. Connect with your volunteers. Say, 'I'm a bee person just like you and I love nature, too.'"

Of course people volunteer for Sam Droege: they like him (I know I do). Part of it is the poetry, of course. Dickinson is a favorite. "Like trains of cars on tracks of plush / I hear the level bee. . . ." Talking with Droege about the poet's ability to universalize observation, which most of us cannot do with words, he commented that the rest of us can participate in making global meaning by contributing data points to a map. For Droege that map is as much a guide to the inner terrain of the observing self as it is a wayfinding to the natural world. You just want to be around people like Droege, do what he is doing, and absorb his sensibility.

Droege is happy to enlist technology's assistance. He's been taking photographs of bees the better to identify them with, by suspending the insects in cuvettes (small see-through tubes) filled with hand sanitizer. Taking multiple images at zoomed-in scales, he stitches the photos together to create a picture of the bee that is so detailed, it can be used by a taxonomist to identify the bee. Droege also figured out how to make the bees ready for their close-up, by shaking them around in a paper towel or using a hair dryer to produce a fetching coiffure. "The ID is the bête noire," he told me. "It's a real limit on what average citizens can do. Identifying bees is super-complicated, and you need to have a trained expertise to do it correctly. I want volunteers to collect bees and send them to me, but this changes the dynamics of the thing."

Projects that don't need this sort of cultivation of a personal connection include eBird. More than one hundred million observations about birds have thus far been submitted to eBird, and let's just say that when numbers get that big they get statistically significant.

While today's burgeoning citizen science movement extends well beyond the avian into similar observational projects devoted to everything from ants to stars to fire to jaguars, undertaken all over the world, bird people started modern citizen science and still lead it. Most projects are grounded in those developed at Cornell. In the 1950s, the lab's founder, Arthur Allen, enjoined local birders to start counting and documenting species while they were out and about with their binoculars. Most birders love lists, they love counting, they love data. Significant numbers started to accrue. In 1994 Rick Bonney at Cornell instigated Project Tanager, enlisting 1,500 volunteers to count four tanager species (sometimes enticing the birds into view by using a taped calling supplied by the lab), collecting data over three years that researchers used to make larger observations about forest cover, fragmentation, and bird health—which all adds up to a better understanding of what human impacts are doing to nature. Other significant research from Cornell's bird projects has to do with tracking avian flu and other aerially transmitted bacterial diseases—stuff that, eventually, humans get.

"But for eBird," Droege considered, "the motivation is self-centered." Yes, while many who contribute to eBird are undoubtedly tuned in to conservation, there is a competitive element that, as Droege said, "really makes the thing work." As you may know from reading Mark Obmascik's fun book *The Big Year* (or having seen the film with Jack Black), avid birders are often consumed with keeping a list of birds sighted, having it be longer than anyone else's, and having everyone else know about it.

Bioblitz

Since many of Droege's citizen science projects lack this inherent motivation, he hit on the idea of "eventizing" an activity that putatively could be done anytime. The idea of the bioblitz was Droege's, but credit for the term itself goes to Susan Purdy, a National Park Service naturalist who helped Droege produce a focused day of counting things in Kenilworth Aquatic Gardens in Washington, DC, in 1996. The obscurity of Kenilworth, one of several small chunks of park property in a low-income area on the wrong side of the proverbial tracks, made it more appealing to Droege, who grew up along the nearby Anacostia River and "spent my youth in the woods" there.

The violent echo of the term *bioblitz*, which refers to the World War II barrage of firepower called the blitzkrieg, has a corollary in the activity's precursor and model, Audubon's Christmas Bird Count, inaugurated by ornithologist Frank Chapman 115 years ago. It was (and perhaps still is) traditional for hunters to stage a competition on December 25. Two sets of shooters down as many birds as possible and then tally up the count. Chapman redirected some of this aggression, and in a twenty-four-hour period on December 25, 1900, twenty-seven people in twenty-five different places in Canada and the United States counted 18,500 birds sorted into eighty-nine species. Every year since, more and more participants have been tallying birds. The 2013 count by 71,531 observers tallied up 67,133,843 birds sorted into 640 species. The 2013 numbers are somewhat lower than average, but the Audubon Web site poses a couple of possible reasons for this, including drought, and the absence of certain high-volume locations from this year's counts. In 2013 the world's tiniest bird, the bee hummingbird, was counted for the first time in Cuba.

Droege told me he conferred with colleagues about how to get people with natural history expertise to do an initial inventory of the park, to see if there was anything interesting in it. Of course, they had no money with which to pay anyone. "So we had to figure out how to lure in participants." Droege thought his idea for a one-day rapid inventory using very simple protocols would appeal to biologists, who "could simply poke around anywhere in the park and use their natural critter- and plant-hunting instincts," without having to do involved documentation. "There would also be other cool '-ologists' there for them to bond with, and plenty of coffee. It was my first experience with eventizing a survey—fixing the date, bringing in media, using social networking to increase participation." Droege's concept took off, and bioblitz popularity has built steadily every year since. The National Park Service now teams up with the National Geographic Society to do a significant blitz once a year in a different state, and attendance numbers and species counts get progressively higher. Bioblitzes using iNaturalist as the data-collection platform are a central organizing feature of NPS centenary celebrations all over the country.

As the three-year project to inventory Mount Tam progressed, both the Marin Municipal Water District and the California Academy of Sciences started to refer to it as a bioblitz, but what we were doing was not really a bioblitz. A bioblitz is generally the carpeting of an area by tens or hundreds or even thousands of people to document every single different species they can find within a designated timespan—usually twelve or twenty-four hours. The MMWD inventory unfolded over the course of three years. Bioblitzes elicit participation from the general public and so did this inventory project, but the citizens who wound up regulars at MMWD turned out to be a highly self-selected group. On many of the teams I've been part of, my teenaged son and I were the only participants without some sort of professional or academic experience with what we were doing. Bioblitzes usually involve taking photographs of species and not actually digging up or otherwise "collecting" specimens, and so book along at a much faster pace. They are also one-off events, not methodical efforts to capture the specifics of a place over seasonal change. (One National Park Service bioblitz is ongoing: the All Taxa Biodiversity Inventory at Great Smoky Mountains National Park—and its stated goal is to discover all life-forms in the park).

(Don't) File It

In a special issue devoted to "Biological Diversity: Discovery, Science, and Management," Jeff Selleck, editor of *Park Science*, comments, "Criteria for what constitutes biodiversity discovery have not been defined precisely. . . . Additionally, most of the work has not been fully documented or analyzed from a national perspective."[5] He's pointing to one of the biggest bugbears in citizen science, which is data collection. What you collect and how are determined by the scientific question being asked, even if it is the very broadest: What species are here right now? Both the Breeding Bird Survey and the North American Bird Phenology Program aggregate data in highly codified, strictly vetted formats and store them in databases accessible to scientists and other interested parties. The data are useful only because of the constraints by which they are collected and maintained.

"There's this whole 'B is for birds' problem," Droege explained to me. "That's a file in a filing cabinet." There are hundreds if not thousands of counts of things like bald eagle nests or snowy owl sightings or anything else you can imagine stuck in manila folders all through the state and federal agencies that have to do with wildlife and natural areas. "Nobody analyzes the data, because they can't," Droege said. It is not just NPS bioblitzes that until recently have been conducted in something of a less than rigorous way from the perspective of data collection, but all inventorying and monitoring done even by professionals. Droege has an "80 percent rule," which is his expectation that only 20 percent of data collected even at national parks by biologists is useful. The remainder is "a waste of time in terms of actual data. People don't do their homework. Usually they've collected data in such a way that it usurps any possibility of getting anything interesting out of it, or they've decapitated the best stuff because they changed the procedures in the middle of the project. Like, 'Let's change from watching sandhill cranes for five minutes to watching them for ten minutes at a time. Isn't it better if we stay longer?'" The answer is no. "Talk to a statistician first," Droege counseled, "before you design your protocol. Then stick to it." He added, "I used to feel really snarky about this but I do realize that, hey, these people are getting out and learning things and that's all good. It's just that it could mean so much more."

Droege pointed to statistics anxiety as the culprit getting in the way of useful citizen science. "Statistics is not about the right answer," he said. "It's about probability. It's a way of thinking about things." Of course our much-lamented national shortcomings in math education have to do with this. But consider also what a sea change it is for us to look at nature from the perspective of "probability," which is about the future. For centuries, we have looked at nature to discern our origins, to re-find Eden, a place from long ago. Now we look even deeper and more accurately into the past, and take sharper inventory of the present, for the sake of finding patterns to help the future as it unfolds.

Today Droege is immersed in a hugely ambitious citizen science project to inventory and monitor native bees in national parks. Few us can have failed to read or hear about the dispiriting massive bee collapse, which is ongoing, real, and scary. The honeybees thus affected are an introduced species harking back to the 1600s, and Gordon Frankie, professor of entomology at UC Berkeley, suspects the cause is probably some combination of stress, pesticides, and monocrops. "We treat them like little machines that you can plug in," he told *Bay Nature* magazine. "They're not happy."[6] Bees also suffer because farmers use them on only one crop. Lack of a diverse diet increases their susceptibility to pathogens, especially viruses. Frankie is looking to encourage farmers to diversify the bee species they use, and also to create some native habitat on the edges of agricultural fields to support native bees.

It turns out we have quite a few native bees on our soil–3,640 described species in North America above the Mexican border, and probably about 400 undescribed species. As elucidated in "Pollinators in Peril? A Multipark Approach to Evaluating Bee Communities in Habitats Vulnerable to Effects from Climate Change," Droege and his colleagues have designed their project to cost as little in time and money as possible. They aspire to aggregate uniformly collected bee data from dozens of parks across the continent. The unprecedented scope and scale of the project will hopefully help us figure out what to do for bees. As Droege's email signatory proclaims: *Apes sunt et non liberum.* Bees are not optional.

One of the interesting things about this project is that it expressly purposes field survey techniques of citizen science to undertake hypothesis testing–which is science with a capital *S*. In this the study design is aided and abetted by the experimental parameters provided by climate change. Droege and company are comparing bee communities in three habitats known to be vulnerable to climate change: high elevation, inland arid, and coastal. These are "representative of the landscape matrix," in this case the entirety of the National Park Service holdings from sea to shining sea, which provides "an ideal natural laboratory in which to investigate large-scale patterns of bee distribution in sensitive habitats and to model how strongly climate change may affect these patterns."[7]

As Droege's paper asserts, collecting the bees is the easy part. Volunteers sample a pair of vulnerable and "common," or more resilient, sites, five times each between early spring flowering and late fall blooming. Thirty "bee bowls" are set out along transects, painted blue, yellow, and white to fool the bees into thinking they are their favored flowers. Trapped in soapy water, the bees are collected after twenty-four hours, labeled, bagged, and sent off for an ID.

Reader, they have to die—imagine trying to ID a bee that is still buzzing. Droege has a handout for those who are concerned about bee mortalities in the cause of science, and its fun facts include these: The average bee has a flight period of five weeks, so bowls capture only a fraction of the tribes of them that come through. Many of those captured are "eusocial," and not reproductive. Reproductive females who die in the bowls mostly have left progeny that are "good to go" without her. "Males do not provide parental care of any kind and thus their captures have little impact on the next generation." Bees die way more often from causes other than bowl traps, like weather, pesticides, parasites, disease, and so forth. And perhaps of most relevance: "Overall mortality of bees from bowls . . . is almost certainly not additive but compensated for through higher births, lower deaths, and immigration. No studies have shown a next year effect from sampling." You could take my word for it that Sam Droege would not hurt a bee if weren't for the bettering of their cause, or you could also check out this discussion of collecting and its impacts: leplog.wordpress.com/2012/12/03/in-defense-of-collecting.

Thirty national parks, preserves, and monuments, ten national lakeshores and seashores, and six national recreational areas, historical parks, and parkways gathered up volunteers who gathered up bees and sent them in to Droege at his Bee Inventorying and Monitoring Lab at Patuxent. "With bees coming in from Alaska to Maine, southern California to Florida, and everywhere in between," Droege reported that the flow was "somewhere between thrilling and completely overwhelming." Especially since Droege took on the responsibility for identifying most of the bees himself. Suffice it to say the job isn't quite done yet—forty-three-thousand-plus bee records takes a bit of time. As of this writing, twenty-five thousand or so have been identified, falling into forty three genera and 685 species. Droege has to send many of the bees out to taxonomists who specialize in bees, and these are "in short supply and in high demand."

The resulting database will constitute the largest replicated survey of native pollinators anywhere in the world. And the study was designed with types of analysis in mind, something that makes it rare among citizen science inventories. Droege and his team will be able to analyze whether the rare or endemic species are in vulnerable or common habitats; they will be able to compare across regions to see if bees on coastal dune sites are more similar to each other than to bees on sites within the same park; and they will analyze elevation, latitude, aspect, soil type, mean air temperature, and precipitation to discern any patterns within and across habitats. We will be getting a map of our world according to bees.

Say Cheese

As adept with the technology as he is with taxonomy and statistics, Droege's most far-reaching proposal thus far, proffered to the rest of us to carry out at the scale necessary, is called Monitor Change. A succinct two-and-a-half-minute video explains it here: monitorchange.org.

"The concept uses little more than a camera phone and a stout piece of bent steel to start," reads the site. Droege figured out that using photo-stitching software and images periodically captured from the same place, he could create a mural of change over time. So, for example, Droege has proposed placing brackets all along the Adirondack trails and challenging hikers to put their smartphone or camera in them while trekking past, take a photograph, and send it in to a central repository (this doesn't exist as of this writing). With the georeferencing utility provided now by virtually every phone and camera, hikers would be providing a piece of scientific evidence of what is going on when and where. Since photo software today can make formats from different cameras uniform, the photos could be stitched together to make a film, essentially, of the real world in action. And make the results available to everybody everywhere anytime.

Droege's idea is being put to use by a sui generis citizen science group in the Bay Area, Nerds for Nature. At monthly meetups in Oakland, this self-described group of "technologists and environmental professionals collaborate to build awesome tools to understand, protect and revive the natural world." There's a drone-building contingent of the Nerds, and group ponderings over how to barcode Bay Area plants and mushrooms. In their emphasis on improvisation and community the Nerds embody the grassroots spirit of citizen science. Two Nerds projects using Droege's camera-bracket idea currently underway are both trained on documenting and observing fire recovery on Mount Diablo in the East Bay, in collaboration with Mount Diablo State Park and the Wildlife Society, and in the Stanislaus National Forest in Yosemite, working with the Forest Service and a nonprofit,

Wholly H2O. If you happen to be hiking in either place, here's what you can do to be a cool Nerd. Find a bracket and take a picture. On Mount Diablo, post it to Twitter using the hashtag #diablofire01. At Yosemite, use the hashtag indicated at each bracket. For example, #firerim01. The Nerds will harvest the photos and "create time-lapse views of change." The effects of fire on the ecosystem here are imperfectly understood, probably subject to climate change, and of the utmost interest to figuring out the deep truth of the landscape, so you will be doing a good deed.

That citizen science in general has big potential for supporting social activism around natural resource use comes into focus, pardon the pun, with Droege's Monitor Change. "The government has put up satellites that show us what's happening to the earth," Droege told me. "The imagery shows, yeah, the climate's warming. But it's at a safe scale." That is, the imagery seems far away, like something we may have to deal with later or at least not right this minute. But citizens can use Monitor Change to create virtual neighborhood watch patrols. Their documentation could help change where development is sited, or help direct restoration efforts where damage is done.

Droege recalled a time in the late eighties and early nineties when the Republican Congress led by Newt Gingrich nearly shut down his USGS bird-recording surveys. Droege told me Gingrich claimed volunteers on the counts were working for Earth First! and falsifying data. Earth First! pushed the envelope on peaceful environmental protest in the early 1980s and earned a reputation in certain quarters as anti-American. Whatever else one might say about Earth First!, the volunteers counting birds and bees for Sam Droege had nothing to do with the organization, and saying so was just the sort of inflammatory stuff we're used to on reactionary talk radio these days. It may seem to the rest of us that an anti-volunteer senator is the true anti-American, and that counting up species to better protect their populations fulfills the spirit of Jeffersonian democracy and stewardship.

Political quagmires, as we all know, pose a significant barricade to getting conservation done. Usually conversations around the government and species-level data collection that aren't about the national parks center on the Endangered Species Act (ESA). The ESA is a citizen science tool par excellence. It was passed in 1973 with a "citizen suit" provision that enables individuals and public interest groups to petition and sue the government to protect species. The act has critical habitat provisions that pave the way for protecting lands and waters that species need to keep on keeping on, and has resulted in protecting hundreds of millions of acres. More than one hundred species listed have since recovered by 90 percent. This is what we want—this is what we need. Just to drum it home again, species are going extinct at a rate and magnitude equal to that which took out the dinosaurs. What do we do about it? Identify which species are most threatened, and protect their habitat. If you protect it, they will come.

Since getting an endangered species designation on your property can mean severely curtailing what other activity is allowed on it, some landowners view the ESA with hostility (plenty more have found ways to coexist with species listings on their land). It's not always slam-dunk easy to figure out the best way to protect species either, and sometimes one species has to be given preference over another. One example is black abalone and sea otters off the coast of Monterey. Both are endangered. The sea otters eat the abalone. Do we stop the sea otters? All over the West it has become clear that restoring periodic burning to the landscape increases ecosystem resilience. But propose burning a landscape where there is an endangered species lurking, and you will be threatened with a lawsuit. This is a case of more science needed, to show that diminished populations of certain species will benefit from fire, even if that population goes down by some number of individuals right after the fire. There's also the question of triage. On some landscapes we may have to let go of one species to benefit a bigger number of species. The continued mysteries of how species interactions really work over historical time frames can be resolved only by getting bigger and better data sets about those species.

So we're back to citizen science, especially as conceived by Sam Droege. In his Los Angeles talk, Droege pointed out that with the advent of smartphones, digital recorders, shared networks, data loggers, and proliferating apps, responsibility for nature "has shifted; now *you* may take control." Participating in citizen science can help effect change in a big way. "The next big picture," he said, will be "the creation of a social network platform that allows many groups to establish networks (public and private) and that either has time-lapse and analysis tools built in, or facilitates such." He pointed out that the time-lapse element allows people to "see" change, and not only is that cool, but the photographs "are the ultimate ticket to credibility" and thus to effecting real change. Photographs are the bridge from "conjecture and anecdote" to "credible, publishable, verifiable data sets." They are data to lore.

As our day hike through Griffith Park came to a close, I told Droege about my nature angst. "It doesn't make sense to me that we are destroying the ground beneath our feet," I said, citing our overconsumption of photosynthesis, to summarize the problem. "We know what we are doing, but we keep doing it." Droege and I discussed the possibility that the extreme intelligence of *Homo sapiens* may be some kind of interim product of evolution that turns out not to be truly adaptive for the species. Droege told me that he hails from generations of Lutheran ministers, and grew up taking our responsibility to do the right thing very seriously indeed. "I had my own collision with what you're talking about," he told me. "I was just frozen with distress about the environment and what we're doing to it. I spent a lot of time doing some Zen retraining of my religious impulses. Now I see it this way: I was taking responsibility for the environment the way Lutheran preachers take responsibility for other people's souls—that's overdoing it. I was arrogant to think I could do something to change the course of what's going on here."

Droege now takes the very, very long view, toward the eventual engulfing of our planet when the sun enters its "red phase" in seven billion years.[8] "Eventually life is going to end," he said. "We are just making that happen sooner." Of course, Droege gets up every morning hatching new plans for putting that expiration date off. In reflecting on his reasoning, I haven't exactly found it consoling. At the same time, he is taking a courageous way forward. I find him heroic in the Joseph Campbell mode. He is the individual bravely going forth, every day bringing necessary information back to his global tribe.

Finding E. O. Wilson

In late March 2015 a conference was convened at UC Berkeley to consider "Science + Parks" in advance of the National Parks Service 2016 centennial. (As mentioned earlier, the NPS was founded at Berkeley). E. O. Wilson opened the show and hundreds gathered in Wheeler Hall to listen to this most eloquent of scientists who seamlessly intertwines cultural and quantitative observations and is unafraid to strike a moral tone. Wilson referenced citizen science as a positive and necessary development in the national parks. Many of the speakers over the next two days would echo the call for citizen science, remarking on the success of the bioblitzes invented by Sam Droege. I mentioned my surprise at this to Dr. Rebecca Johnson, as referenced earlier, a leader of citizen science efforts at the California Academy of Sciences. Johnson is a PhD of a younger generation than the mostly male gray-beards coming to the dais.

"It would seem a lot of hard-core scientists are coming around to citizen science," I remarked.

"It's like they have to hear it from each other," she said.

Wilson's whole talk was good and his ringing refrain focused on extinction. Wilson asserted that it is more important to our future life to protect plants and animals from extinction than it is to focus exclusively on excessive emissions. "If we save the living environment of the earth," he said, "we'll also save the physical or nonliving part of the environment, because each depends intimately upon the other." He said it is downright wrong to imagine "that later generations will somehow find a way to equilibrate the land, sea, and air in the biosphere on which we absolutely depend." (The "biosphere" is a comprehensive term including the geological, hydrological, atmospheric, and biological carbon cycles—basically it describes the toto picture of Earth as a living planet.)

Wilson's prescription is a number: 50 percent. He proposes that as human development proceeds apace, we limit our ingress and leave half of the biosphere to nature's own doings. It doesn't work, of course, just to save any old half for nature. The connectivity concept as borne out by William Newmark's work in the national parks will need to be applied. Wilson advises that we save the mega linkages first—for example, what's known as the spine of the continent, which runs the length of North America from the Yukon to Mexico. Thus plants and animals will be able to move as the climate changes, and they will be able to mix with others of their own kind from different populations. This is not nature cordoned off away from humans. Wilson envisions a vast interconnection of wilderness areas interpenetrating with human development. Conservation organizations already working to achieve large landscape connectivity include the Yellowstone to Yukon Conservation Initiative and the Wildlands Network. In 2009 Wilson's "nature needs half" idea was developed into an initiative launched at WILD9, the Ninth World Wilderness Congress, held in Merida, Mexico.

In "Nature Needs Half: A Necessary and Hopeful New Agenda for Protected Areas," Harvey Locke noted that until now we've defined conservation targets in a most ill-defined way. We have "set goals that are politically determined, with arbitrary percentages (of land and species) that rest on an unarticulated hope that such non-scientific goals are a good first step toward some undefined better future outcome."[9] By not literally quantifying how much aquatic and terrestrial real estate is necessary to meet the quantified needs of healthy biodiversity, we have let far too much slip through nature's hands and into the roiling maw of development.

Wilson delivered his talk starting at 8:30 AM and a string of speakers followed him. The conference was due to break for lunch at noon and to reconvene at 1:30. At 11:45 I felt like taking a walk in sunny Berkeley rather than joining friends for lunch; to avoid making excuses I decided to slip out of the auditorium early. I entered the large, mostly empty foyer of Wheeler Hall. There was a lone man sitting in a lone chair looking over some papers. It was E. O. Wilson. This was like coming across Bono by himself at a U2 concert. There were hundreds of biologists in the next room!

As mentioned earlier, Wilson and I have communicated in the past but I had never met him face to face. I walked over and introduced myself. The Southern gentleman stood. He is tall and leaned forward to tell me he was delighted to see me. Quickly I told him that I planned on explaining in my forthcoming book how we can help achieve nature's half by using citizen science. Wilson knew my book about the spine of the continent, a mega linkage such as he advocates protecting first. I had (and have) the idea that citizen science could sidestep many of the political and emotional hurdles to saving nature along this continental scale swath of nature. I looked over Wilson's shoulder, anticipating the handlers who were no doubt going to separate me from him any minute.

My general idea, I explained, was that we capitalize on already-vibrant centers of conservation activity, like the Sky Island Alliance in Tucson, Wild Utah in Salt Lake City, and the Denver Museum of Nature and Science, and connect these via a Web site. On this Web site would be visualized the data collection of citizen scientists trained on a variety of taxa and natural phenomena on local landscapes. Those using the Web site could begin to make connections between, say, snowmelt in Colorado and a higher or lower abundance of pika, the little rabbit relatives who live at very high elevations in Montana. "Citizen science is good right now at helping people collect data," I told him. "But the next step is to give them a way to analyze the patterns and put two and two together. This isn't outrageously expensive." In fact, it's a matter of a few staff people to organize things, and some good mappers, and, importantly, a talented designer or two. "We make a portal," I said, "so that people with local development issues can aggregate information about species and landscape impacts—to show people, hey, you want to put a parking lot here but look who's using this place, look who lives here."

Wilson has one droopy eyelid. He looked at me intently through both eyes but of course I spoke directly to the eye that is more revealed. "We add up the species for people and have little lessons about population dynamics so they can see the extinction pressures mounting."

He was smiling.

"Because people don't understand extinction," I said.

"No, they don't!" he heartily agreed. "Even scientists don't understand it."

"And this is the most important issue, not of our time," I was emboldened to declare, "but of any time."

Wilson took my arm. "I'm going to help you," he said, and he introduced me to his now-hovering assistant. In moments Wilson was spirited away in a dashing display of gubernatorial theater. Jerry Brown had arrived with an entourage of state troopers who cleared the area. I went for my walk. I heard later that Brown and Wilson talked together for an hour and a half, surrounded by a circle of law enforcement keeping the rest of us at bay.

CHAPTER ELEVEN

Eyewitness

Eight-thirty AM was roll call for hawk-watch duties at Fort Cronkhite; at this hour it took me less than ten minutes to drive from my home across the Golden Gate Bridge and hence to a quick exit toward the Marin Headlands. It took ten more minutes to drive to the old barracks where Allen Fish, director of the Golden Gate Raptor Observatory (GGRO), would present the day's marching orders, and the road wound through an already transposed world. From the built city one was now among tall brown hills, folded and faulted, their crevasses sporadically sprouting dark green cypress canopies. Sometimes I rode my bike to the Hawkwatch and the slower pace allowed me to discern many a figure in the carpet—bobcats, coyotes, great horned owls.[1] This spit of land pointed at the juncture between ocean and bay; waves crashed up and around bare rocks, light streaked down over the water and hills through restless clouds, and much of the view could be isolated from any sign of human presence. Timelessness was an illusion easily penetrated, however; on my first day of hawk-watching in September 2012, a somber processional made its way across short, chert-pebbled Rodeo Beach: six emergency medical technicians, two police officers, and a gurney with a white-shrouded body on it. This was a case of foreshadowing in the narrative of me on the Hawkwatch.

After the debriefing in the barracks, we headed up to Hawk Hill, about a five-minute drive, and took our positions in a runic circle facing the four cardinal directions, each group of two to three sharing a telescope and all of us attached to binoculars. The hour would be called out and we would switch positions, sometimes mixing up the groups, sometimes staying with the same partners all day. North was the best; here was where most of the hawks emerged, heading south. In this direction you also got to stare at Mount Tam, glowering and squat, making its own weather over there. West there was Point Bonita between the Pacific and the Bay; on a super-clear day you could see the Farallon Islands, where hundreds of thousands of seabirds touched down each year to breed. Circling south, there were porpoises, seals, and often whales in the water, with gigantic cargo ships gliding past. The Presidio looked like a green shrug around the city's shoulders. If you squinted just a bit, San Francisco looked like a revealed geologic formation of sandstone or a cubist canvas. East, there was Angel Island and the Golden Gate Bridge, over which the birds disappeared. "They work up the nerve to cross the water," said Allen Fish. "When they finally get to Mexico they pull out their Michener novels at the beach."

Several months into the Hawkwatch I understood perhaps a little too well that I was not going to successfully identify a hawk anytime soon. "Two-year apprenticeship" in fact sounded kind of optimistic. In the shorthand way of calling out bird sightings by landmark, I could say, "Bird over Golf Ball," or, "What's that under the North Tower?" but just when "Turkey vultures over Angel Island!" started to come up in my throat, someone quietly noted, "Bunch of ravens over Angel Island," more as an aside than anything else. Oh.

"Why aren't you guys interested in ravens?" I asked no one in particular one day, and about five people answered: "I love ravens!" But ravens were common and lived here; they were not passing through on their way south, avoiding the Sierras, taking advantage of the ridge and peninsula status of Hawk Hill from which to scout dinner and capitalize on updrafts, flying across the bay at its narrowest opening, which was why the Golden Gate Bridge was placed there as well. Ravens were not noted but up to eight genera and nineteen species of migrating raptors a day were, usually down to the age (juvenile or adult) and the sex of the bird. The numbers and details of hawks migrating along this flyway have been documented since 1985—one of the longest and most successful citizen science projects in the country. Data collected over this long time frame informs research on bird distributions, the spread of disease, and the impacts of climate change, among other things.

The GGRO Hawkwatch is organized on a fourteen-day cycle from August through November each year, with more than three hundred total volunteers committed to a full day of hawk-watching or banding once every two weeks. Since 2012, with one year out to join the banders, I have been on "Wednesday-one" of the watch. Our debriefings include information about the travels and whereabouts of any birds being radio-tracked, species counts to date, and in the case of 2012's Hawkwatch, further reporting and wonderment at a surge of broad-winged hawks (genus: *Buteo*—I'm studying the books as well as the sky, and doing better at it). On September 27, 295 broad-winged hawks were tallied in the six-hour stint of the day's count, ten times the previous one-day count for broad-wings, and exceeding the GGRO's entire season record by ten birds. By October 8, the broad-winged count was up to 725. Why broad-wings all of a sudden?

"I really don't know," said Fish, relishing not just the novelty but the poetry of this happening. The discovery that Hawk Hill is the site of a major raptor migration was instigated by a "crazy lady" who in the early 1970s kept reporting the seasonal presence of an East Coast species, the broad-winged hawk, around her home in Twin Peaks to Laurence Binford, then curator of ornithology at the California Academy of Sciences. Binford didn't believe the broad-winged part of her story but wondered himself where the autumn hawks outside his office in Golden Gate Park were coming from. He investigated. In 1979, Binford published a technical account of the raptor activity at Hawk Hill (he called it Mount Diablo) in *Western Birds*. Over six seasons he logged 263 "sporadic" hours and counted eighteen species (including the broad-wing) of nearly 8,700 hawks, and thus the landmark was established.

Given that my actual counting contribution was nil and likely to remain so for at least another season, while I kept my eyes on the sky at Hawk Hill, my mind did wander. If I trained my attention on a patch of sky long enough, a bird would appear, but where did it come from? The birds traveled through the atmosphere like they were needles darning through an opaque fabric suddenly become translucent. They disappeared the same way. As we rotated observing position around the quadrant, the diversity and magnitude of natural beauty made a clarion call to heart and soul and challenged a brain trying to synthesize the details, qualify the experience. The world from up here was all sky, water, and hills; the busy human imprint of sailboats and even skyscrapers were incidental decorations. Aha! Yes, this was all much, much more than I could comprehend, and that was cool, I accepted my part in it. I looked around: Was this why these people were here?

A couple of Wednesday-one watchers were interns at various stages of environmental educations and so the reason for their participation was self-evident. But others had logged a decade at the Hawkwatch, some twice that time. "Do you like the hawks in particular because they glide and soar like this in the open sky, where you can watch them longer?" I asked Christine Cariño, the "day leader" of Wednesday-one. More than a decade ago, Cariño started as a bander with GGRO, then took a couple of years off while homeschooling one of her children, who has a double classification of learning difference plus exceptional intelligence. She resumed as a hawk watcher several years ago. "No," she said. "Hawks are top predators. They're killing machines. There's just something about that." Cariño is a soft-faced woman who usually has whole grains for lunch.

Similarly you might wonder from whence issues Fish's inclination toward gangster birds. Photographs featuring Fish, a mushy bear of a man, in biannual "raptor reports" over his thirty-one-year tenure with the program show a handsome, bright-eyed guy becoming avuncular. He has a thing for Jungian psychology and once made a film with Gary Snyder, who, he said, "really gets it, that the Hawkwatch is a bioregional activity, it's about taking a deeper stake in where you live." Fish repeatedly cautions against arrogance in making IDs and enjoins even his most seasoned watchers to check with someone else before resting with certainty that the bird just called out is truly a harrier or a merlin. One morning's debriefing was spent on the protocols for "passing." This is when watchers facing one direction see a bird or birds but aren't sure what it is, so call to those stationed at the direction it's flying toward. "Remember you are asking for help with verification. Don't shout 'adult red-shouldered heading east' like you are slapping the folks next to you with your guess. Be humble." Fish is likewise humble about quantifying exactly what we are counting here—the birds come and go, disappearing and appearing. Heraclitus said you can't step in the same river twice but you can certainly count the same bird twice. Fish settled on the four-directional quadrant to capture directionality as well as sheer numbers of birds, and these amounts he translates into "overall raptor activity" rather than a definitive quantity of feathers and beaks.

Occasionally I have dropped in on other groups on other days, partly because all the senior Hawkwatchers have personal, idiosyncratic, and fascinating knowledge of their subject. The quality eluding many citizen science projects—a sense of community together with a charismatic leader whose natural knowledge and willingness to share it are constant motivation—is on overflow here at the Hawkwatch.

There is no doubt that Allen Fish's GGRO Hawkwatch is a community phenomenon and includes birds as our neighbors and cohorts. As a citizen science program, it's hard to beat, hitting every single objective of such with square-on satisfaction for the participants, and yielding data that is relevant far beyond the snug base of Fort Cronkhite. A large contribution to what makes the Hawkwatch work so well is in fact an old-fashioned artifact: the *Pacific Raptor Report*. This handsome publication is printed in small numbers and disseminated widely online. It reports out on hawks counted and banded by species and date, and provides updates on current research purposing the data. It profiles volunteers and acknowledges their deaths. It includes opinion pieces and instructive musings, such as Buzz Hull's 2006 article "Why Do We Care What Age That Red-Tail Is?" In confabs discussing how to mount and unroll monitoring and inventorying projects, you will almost always hear tell of the need to present results back to participants and solicit feedback. This isn't a fancy digital age concept. It's called a newsletter.

The Hawkwatch always feels like it is bigger than the sum of its parts. Since it goes on seven days a week for four months, on any day August to November you can look over at Hawk Hill, above the Golden Gate Bridge, and know that at least eight people, and usually about fifteen, have their eyes glued to the skies, as if their attention were holding the whole scene in place.

Harbinger

From the dove Noah spotted that signaled land to Odin's twin ravens Thought and Memory, people have looked to birds for information not otherwise evident to human perception. Bird listing, occupation of the avid, has produced volumes of data referenced earlier, on the Audubon Society's Christmas Bird Count and through various Cornell eBird programs. Population dynamics research has capitalized on this data bonanza since the very early days of the Bird Count, but as the numbers accrued, a pattern was emerging, and it represents perhaps the most consequential message birds have yet had for man.

In the early 1980s, Terry Root was working as a scientific programmer on the *Voyager* spacecraft, crunching numbers for a cosmic ray experiment at the University of Colorado. "I was also a birder," she said, and very good at computers. "Someone told me there was a professor at the university who had a mag tape and didn't know what to do with it." The professor Root heard about was Carl Bock, an expert on fire disturbances and birds, and he had an idea the Audubon data should be purposed to greater effect than it had been. They gave it to him, but that was as far as he got. "I played with the data," said Root. The rest of us can sympathize with Bock—that's a lot of numbers on some staggering spreadsheets.

Root is a tall, liquid-eyed beauty with a solid sheaf of gray hair to her shoulders. Her demeanor says senior Stanford researcher, which was her job until her recent retirement, but it is hard to resist seeing in her face a certain sensitivity and persistent sadness. Root is the widow of Stephen Schneider, a legendary climate scientist, and also a huge citizen science advocate, who died suddenly in 2010.[2] The term *citizen scientist* is usually credited to Rick Bonney, but Schneider used it with a political edge, as when he declared it time to "empower Americans to become citizen scientists" at the 1997 meeting of the American Association for the Advancement of Science. The global warming Schneider insisted was happening, against virulent resistance and opposition, was signaled by Root's work on the Christmas Bird Count data. In order to make informed decisions about climate change and other environmental threats (he referenced improperly stored nuclear waste), Schneider said people should learn to ask three questions of the scientific community: What can happen; what are the odds; and how do you know?

Root's *Atlas of Wintering North American Birds: An Analysis of Christmas Bird Count Data* (1988) is among the earliest applications of citizen-gathered information on the natural world. The book is about three hundred pages of maps of the lower forty-eight with squiggly demarcations and crosshatches customizing each to express where hundreds of species of birds go in the winter. Root conceptualized a natural phenomenon in a new way and not everybody liked it. She proudly sent her work to Roger Tory Peterson, famous for his field guides, and he dismissed it. Computers can't tell you where birds are, he told her, only people can. Paradoxically he affirmed the citizen part of her project while dissing the "scientific" part.

"One of the reasons the data appealed to me so much is that it was far easier to sit behind a computer all day than go out in the field with the guys," Root told me. While she is well-known for research and publications on bird distributions, Root has another subject she'd like to write about someday: the situation of women in science. When I asked her about that she told me she has been collecting notes and writing them up here and there, but is still reluctant to publish them. She'll talk about it though. Bock turned out to be a boon for her. "His wife was on the faculty. He didn't expect me to sleep with him. All was well."

In the mideighties, the era Root referenced, the field of conservation biology was absolutely brand spanking new. Ecology had been studied for a long time, yes, but it was not until 1986 that genetics, molecular biology, population studies, and a host of other disciplines were braided into the study of what is going on out there with the express purpose of figuring out what is going wrong. Virtually every woman over forty whom I have interviewed for this and other books who was working in conservation biology in the 1980s has told me, off the record and on, that the field was exuberantly rife with what is now called sexual harassment. Several of them chose graduate programs based on where they might be preyed upon least.

But I guess in this case we can thank the satyrs for their leers. Root's time behind the computer screen not only paved the way for citizen science in general to be taken seriously, she also helped forge its first connection with climate change. "At that time, biotic interactions were driving the study of ecology," Root told me. *Biotic interactions* are the manners, mores, and survival practices of species. The term includes predation—who eats whom—and competition—who wants the same thing as someone else, and how they go about getting it. Root pointed out that in 1986 Peter Kareiva, a mathematical ecologist who has recently become director of UCLA's Institute of the Environment and Sustainability, published a paper observing that more than 80 percent of the study areas looked at by ecologists were the size of two tennis courts put together. She changed the scale of ecological inquiry.

The vast numbers at Root's disposal and the continental reach of where they came from allowed her to look at what was going on from a different perspective. "At the small scale, yes, you see predation and competition are important. But you can't see what temperature has to do with things, or the impact of varying vegetation types." Vegetation is closely linked with precipitation patterns and temperature, the primary measures by which climate change is assessed. "The large-scale issues opened up questions that couldn't be asked any other way." In sum, her analysis showed bird distributions moving decidedly north (species are moving poleward in response to the changing climate; these are North American birds), and this correlates to spring arrival on average ten days earlier than historically. Her analysis has been thoroughly borne out by many other ways of measuring the same phenomenon.

All these numbers add up to a very good picture of how bird populations are doing everywhere, of interest not only in itself but as an indicator about how the world's environment is doing in general. Depending on the species, bird vitality has to do with the place where you spot them and also with the places they are coming from and going to. Birds are a significant weft upon which biodiversity is woven, and when this thread is fraying or snapped, you can be sure the fabric of life is in bigger trouble somewhere. If we can't quite say how nature works, we can define its attributes, and one way of measuring nature is by "ecosystem services." As articulated by the UN Millennium Ecosystem Assessment, birds provide all four central ecosystem services: they provision (provide a food source), they regulate (for example, by controlling insect populations that would otherwise explode), they are culturally important (see Odin and Noah), and they provide supporting services, like pollination.

Migrating birds add a spatial and temporal dimension to all this and link ecosystem processes separated by great distances.

While general opinions are changing, the scientific community is not uniformly keen on citizen science, and it questions whether mob-sourced data collection can possibly be relied on. Terry Root is an ardent supporter of citizen science, traceable perhaps not only to her singularly important contribution to the phenomenon but also to her willingness to conceive science as a wider-open field than the old boys would have it. Fundamentally, though, Root said, "Real-world problems are so large now, we scientists can't do it ourselves."

Chrysalis Crunch

At the same time Root published her climate change research based on birds, Camille Parmesan came to the same conclusion tracking butterflies. As with Paul Ehrlich and Peter Raven's butterfly research (and Parmesan is one of Ehrlich's butterfly progeny), the basic data set of collected butterflies is at least as much the work of amateur collectors as professional scientists—and the older specimens are almost all collected by crazed hobbyists. Parmesan set off with a NASA fellowship to count, first spending a year visiting natural history collections in the United States, Canada, Britain, and France, tallying up where her study subject, Edith's checkerspot butterfly, had been historically collected, and when. She spent six more months tallying up pinned specimens in private collections. Her list of locations where Edith's checkerspots had been historically sighted exceeded 1,400. Of these, more than six hundred records of sightings were detailed enough that she could find them again, and that's what she set out to do. Hopping in a four-wheel-drive pickup truck, she traversed North America from Mexico to Canada, traveling more than forty thousand miles and visiting 292 of the documented sites where Edith's checkerspot had historically been known to tarry.

Parmesan, now a professor at the University of London, found that Edith's checkerspots "were shifting their entire range over the past century northward and upward, which is the simplest possible link you could have with warming. . . . What I got was 80 percent of the populations in Mexico and the Southern California populations were extinct, even though their habitats still looked perfectly fine."[3] Parmesan published her findings in a 1999 paper in *Nature*. "The biological community was thrilled," she said, "because of the amount of data I had and the scale I was looking at—the entire geographical range," from Mexico to Canada, of Edith's checkerspot butterfly. Parmesan added that climate scientists were thrilled too, because they were detecting a warming trend but this data showed it was real—what they were seeing was having an impact on biology.

It sounds kind of simple: take note of where butterflies have lived and see if they still live there. Edith's checkerspot is in many ways a model species for tracking response to climate change, since it is very temperature-sensitive, and doesn't move much. Monarch butterflies migrate, of course, but Ediths are stay-at-homes, more or less, and spend their lives within an area of several hundred feet. Complicating matters is the fact that checkerspot populations routinely "blink out." This was the case in the famous study I talked about earlier, in which, under the watchful eye of none other than Paul Ehrlich, the Jasper Ridge population of Bay checkerspots disappeared in Palo Alto. The example is a textbook illustration of island biogeographic theory—Jasper Ridge became an isolated island of habitat too far away from a source of renewal to keep the butterflies resident.

Parmesan became well known in scientific circles and was named one of only four biologists (with Root) to consult on the first Intergovernmental Panel on Climate Change (IPCC). To develop a report on how climate change is impacting biodiversity, Parmesan joined up with Gary Yohe, also on the IPCC. Yohe, a professor at Wesleyan University, is an economist who specializes in adaptation and mitigation impacts of climate change; in many ways, Yohe, as an economist, speaks a different language than Parmesan, as a biologist, does. At first they argued a lot. Yohe was concerned that most local changes in populations, measured over the short term, are due to a natural ebb and flow and to land-use changes, like the development that took out the Jasper Ridge checkerspots. Yohe wanted to make sure he and Parmesan put together an analysis of climate impacts on biodiversity that separated these impacts from those resulting from climate change. Parmesan told me she and Yohe had to work hard to find a common language with which to collaborate. On the one hand she had to teach him some simple biology, and on the other hand he had to get her to look beyond ivory tower science.

Eventually, Parmesan hit on a plan, to do a meta-analysis, incorporating multiple species. She and Yoke correlated springtime appearance changes with temperature, looking for what they termed a "sign shifting," or a change in the pattern over time. What they did is expressly what citizen science data collected by Nature's Notebook is used to do. By tracking changes in nature's events like flowering and migration, and correlating these with temperature changes, a pattern of climate change influence can be discerned.

Parmesan and Yohe thus focused on phenological or timing shifts—when critters come out of hibernation, when the first bud of spring opens—and range boundary shifts such as she had documented with the Edith's checkerspot. Yohe's ambition to determine what economists call global coherence—which means a process or event can be quantified as having a similar effect across multiple locations around the globe—led them to zero in on something Sherlock Holmes might appreciate, a "fingerprint" identifying climate change as the driver of fundamental changes in the way species are behaving. And Parmesan's findings were dramatic: 65 percent of species, not just butterflies, had "jumped their historic ranges," correlated, again, with temperature increases. Their coauthored paper on the subject, "A Globally Coherent Fingerprint of Climate Change Impacts Across Natural Systems," was published in *Nature* in 2003. It now approaches *The Theory of Island Biogeography* in frequency of citation by other scientists.

As citizen scientists provided the raw materials for Parmesan's studies—essentially centuries of data points about butterflies—so citizen scientists will be ever more important for updating her work. We need not only to know that climate change is happening, but to document whether it is going faster or slower than we thought. We need to track its impacts—and species like butterflies illustrate responses by where and when you find them.

Two If by Sea

In August 2013, Parmesan collaborated on a paper likely to have a similar impact as a reference point for future science. The primary author of "Global Imprint of Climate Change on Marine Life" is Elvira S. Poloczanska, and there are a total of twenty scientists signing off on this one, part of the Fifth Assessment Report of the IPCC. "I'm a terrestrial scientist but I was brought in on the work because of the meta-analyses I've done for land animals," Parmesan told me. It's not quite as straightforward to count up the denizens of the deep as it is to count butterflies, "but you can still take censuses. Polar bears are marine species—we have census data from airplane counts." One amazing (citizen science) data set comes from the Sir Alister Hardy Foundation for Ocean Science. "They built devices to be dragged behind ships that automatically take water samples. They convinced commercial ships to drag these and have now accumulated a forty-year database with GPS recordings on a daily basis!" These samples measure the density of phytoplankton and zooplankton. And from the tiniest of ocean life to the largest, again, there is a uniform finding: species are moving poleward as the climate changes.

"The ocean isn't warming as fast as the land," Parmesan said, "so the thought was the impacts wouldn't be as marked." In fact the study shows marine life is having a much stronger response than its terrestrial brethren. "Putting the terrestrial and the marine data together, we now have more than four thousand species in our studies," said Parmesan. The terrestrial species are moving about four miles per decade, the marine species about forty-five miles per decade. Are these plants and animals going to be able to outrun climate change? "We are not seeing any population of species able to live where they have not been able to live before," said Parmesan. "They are playing around with the range of where they have lived. It's really difficult for a species to shift to a new climate space—this happens over millions of years." Parmesan doesn't see evidence that we have crossed a point of no return for most species. "But every year we don't do anything it gets harder."

Cup o' d-CON

Frequently Allen Fish placed one of his collection of raptor puppets unceremoniously on a tree stump up on the Hill; this was, yes, for the amusement of his volunteers, but also to soften the visual effect of a group of people standing stock-still for six hours in a circle when tourists clambered up to see what on earth we were doing. The role Fish takes as emissary of the

Hawkwatch to all and sundry who ask is something the Golden Gate National Parks Conservancy and the Golden Gate National Recreation Area people who fund him find extremely valuable. Essentially Fish presents an outdoor laboratory and a hands-on experience—lending his binoculars and proffering a telescope— to thousands. (More than thirteen million people visit the Golden Gate Recreation Area a year. One of the largest urban parks in the world, GGRA includes Muir Woods, Alcatraz, the Presidio, Golden Gate Park, and many other highly visited destinations, though the Marin Headlands are right up there in popularity.) Many of those who make the pulse-pumping trek up Hawk Hill are birders from around the world who have put it on their list of must-sees in the Bay Area.

"That owl looks sleepy," said one of the super-senior Hawk-watchers, nodding at a cozy little puppet nodding on one of the stray pylons left over from some military structure or other and still stuck into the ground. Actual owls are spotted with frequency at Hawk Hill, mostly by the highly experienced, who somehow discern those staring eyes from the camouflage the animal effects in the vegetation. "Naw," said Fish of the somnolent owl puppet, "it's had a cup of coffee." "Not Peet's!" someone lamented, and this light moment became somber.

Peet's Coffee has long been more than a cup of good coffee in the Bay Area. As a brand that started out as a community effort of tasters in Berkeley, under the guidance of Alfred Peet, the coffee has typified the self-concept of many a local denizen. It has stood for our equable way of going about things, our excellent general good taste; as a strong local company employing people who live right here, it epitomizes our ethic, our pride. Watching Peet's go national has been okay—after all, we want the company to succeed, and we are happy to drink

Peet's at JFK airport. But in July 2012, the board of Peet's had sold itself to Joh. A. Benckiser Holding Company, a 10 percent owner of Reckitt Benckiser, worth $40 billion, and which among other things makes d-CON, an anticoagulant rodenticide. The EPA has been trying to get d-CON off the market for years, and Reckitt Benckiser has "thumbed their noses at them," said Fish. d-CON kills rats, all right, and also dogs and cats that eat the rats, and raptors. A mate and several progeny of Pale Male, the red-tailed hawk immortalized in Marie Winn's wonderful book *Red-Tails in Love*, are among d-CON's casualties. Allen Fish is simply heartbroken about this turn of events. He has been a voluble appreciator of Peet's coffee and he doesn't want to have to change his brand. With Lisa Owens Viani, Fish founded a group called Raptors Are the Solution (RATS). Its efforts to combat the rodenticide include trying to get Peet's to confront its new parent about d-CON.

In a very unhappy way, the Peet's-d-CON conjunction is an instance of *plus ça change, plus c'est la même chose*. As dying hawks are today a message about poison percolating through the natural systems we would like to pretend don't exist, so shrinking hawk numbers alerted the twentieth-century ecologist Rachel Carson (another model citizen scientist) to the perils of DDT. Carson formulated the trajectory she limned in *Silent Spring* in part by reading hawk-counting data such as that being accumulated at Hawk Hill. Consulting tabulations kept by Maurice Broun at the Hawk Mountain Sanctuary in New Jersey, she noticed that from 1935 to 1939, 40 percent of the eagles observed were yearlings, but by 1955, these dark-plumed youngsters had become rare. In 1957 there was only one young eagle per every thirty-two adults. Broun tabulated the numbers; Carson recognized the population crash. She also traced its cause. Based on

Carson's work, the general public made a collective decision about DDT and it was banned; that is citizen science in Steve Schneider's definition.

One day in early November 2012, Fish reported on meeting with the CEO of Peet's, who had pledged fealty to the Bay Area community and wanted to help, but prevaricated about what, exactly, to do about the d-CON. *So* awkward to actually get the head of Reckitt Benckiser on the phone about it. Fish was inclined to wait and see if the CEO would make good on promises to help RATS in grassroots efforts to raise consumer awareness of what is going on; Fish's colleague at the helm of RATS didn't have his faith, or his patience, and wanted to start a big boycott of Peet's. While he hasn't taken that big step, Fish told me, "I enjoyed my last cup of Peet's with the CEO, while he said, 'Nothing will change at Peet's.'"

F-16s over Slacker

It's called Fleet Week but it's really Fleet Weekend, and on Friday, October 5, 2013, I made the ascent to Hawk Hill on my bicycle to join the Friday-two watch. It was a gloriously clear day and the bird numbers had been high all week. I knew from experience that although the day's scheduled air show didn't include the Blue Angels, who had rehearsed the day before and would officially perform on Saturday and Sunday afternoons, they would in all likelihood slice up the sky at approximately 1:30 PM. I was curious about how the hawks would deal with it.

Beginning at around noon historic airplanes started performing in the sky. Here was one that looked like a bus; here was one that might have Snoopy's Red Baron in the cockpit. Today there was visual competition with the doings above: seventy-two-foot international catamarans competing to qualify for America's Cup below. The dun-colored wing-sails were of course as high-tech as had been conceived up to that very minute, but they looked like prehistoric feathers, enormous and kind of ugly in contrast to the pretty little white sails dancing around them in the Bay. A couple of days later Hawkwatchers would witness an American AC72–mast height 131 feet–flip in the water, its tiny crew hanging on for a moment then dropping off. "One second later the ebb tide dragged it under the Golden Gate Bridge," a Hawkwatcher told me. "They had a helluva time getting it back."

The whole city seemed to be throbbing. The 49ers were playing; the Giants were playing (and would go on to win the World Series the next week); Warren Hellman's free concert, Hardly Strictly Bluegrass, was going on without him for the first time in Golden Gate Park. "Six turkey vultures south," called a Hawkwatcher. I turned my binoculars over to Ocean Beach, where at first I saw nothing but sky. Then, there they were, six silhouettes, but they were not birds, they were planes. Love them or hate them (the sound is rending and this showing-off costs millions of dollars), my preference was to succumb to this exciting display of martial power. The tight formation of the six planes, the clarity with which they split off from each other and then realigned, was something like what birds do but not. One of their favorite tricks is for two planes to peel off and leave the remaining four to zip here and there, now all in a line, now stacked on top of each other, while your eye calibrates the inconceivable uniformity of the short distance between them– *how do they do this?*–and then disappear. You have forgotten the other two planes, but here they come, straight at each other, and zinging backward away from a collision like ravens playing in an updraft.

The hawks, by the way, were undeterred. "Adult coop over Slacker" referenced one of the landmarks we use to locate our sightings. The birds we saw flew much lower than the planes. It was possible that birds flying higher than our telescopes could discern were waiting this thing out. A squadron of pelicans flew by. Their coordinated ease and majesty upstaged the fighter jets.

The spiritual symmetry between the hawks and the Blue Angels and the military past of the Headlands—we had piled our snacks and backpacks up on a leftover concrete radar base—effortlessly integrated natural and human history and the overall pleasant feeling was, here we are in time, and aren't we lucky. The man-made elements here, the bridge, the sailboats, actually added to the experience. A current dustup in conservation discourse is over whether we should restore natural landscapes to a "pristine" image of the past, or whether we should accept that nature is degraded and just get on with it. This is something of a straw issue. I have met not a single conservation biologist who operates with "pristine" in mind. They don't want nature to be perfect, they want it to function. Healthy functioning has emotional and cultural components. Fish told me about his anger and angst that, "as species are lost from native ecosystems, we also lose hundreds if not thousands of stories about the intricate relationships that have evolved between those species and their neighbors." What will the story of Noah's Ark mean if we lose lions from the ecosystem? That dove with an olive branch in its beak?

Now that I am a (still green) veteran of the Hawkwatch, I admit that many of the initial frustrations of learning the hawks do persist, but so do the pleasures of the quest. These were neatly encapsulated on November 28, 2012, "Data Be Damned Day." This was the final Wednesday-one of my first season, though teams were due to keep counting through December 9. The plan to bring champagne and more snacks than usual up to the Hill was diverted somewhat by the auspicious opening to what was to be three days of storming rainfall, and we lingered at Fort

Cronkhite, taking a raptor-identification test that would have been humiliating if anyone insisted on looking at my results. ("Jill, lock the doors," said Fish as he revved up the projector.) Since my score was a foregone zero, I utilized the exercise by trying, in fact, to learn from it. So instead of guessing at the name of each of fifteen birds, I described them on my test form: "White body, streaking on breast, narrow, tight tail, fingered feathers on wing tips" (a broad-winged); "Big white spot at juncture of tail, close to the ground, seems small" (harrier).

Though she got most of them right, a four-year watcher was disheartened by her imperfect score. "You have Dennis, you have Margaret, you are getting perfect identifications from these people. I'm not getting any better. What am I doing here?"

"Mimosas on a rainy November day in the Marin Headlands!" said Fish, laughing, but there is no scientist more ready to extoll the utility of imperfection. Fish told us about a hawk counter at Cape May, the biggest raptor migration site in the country. "His name was Frank Nicoletti, and he had some kind of special visual acuity. When he counted he pushed the radius of sightings out much, much farther from what anyone else could see." The number of hawks accounted for increased significantly. "But they had to fire him after two years," Fish said. "One person can't make such a big difference, or the program isn't replicable. There's no way to track relative abundance from year to year if one set of eyes changes the count that much." Fish also reminded everyone that even Dennis and Margaret made mistakes. "The biggest meeting point with all of you is the pure love of seeing a raptor fly over Hawk Hill. To see a red tail illuminated by late September light. The importance here is the hawks and *you seeing the hawks*."

As I did with the tide pool monitoring, I regularly reported Hawkwatch doings to my father. "It's frustrating that I can't identify them better by now," I confessed. "But in a way it puts the activity squarely in the realm of witness. Allen is always telling us, 'It's about you observing the hawks.' And so they go by," I told him. "I'm just observing."

My father quickened to this. "That's spiritual," he said.

The Hawkwatch is a seasonal event, unlike Beach Watch, for example, which is a year-round inventory of what's happening on the shores. Unlike the inventory of Mount Tam, the Hawkwatch counts not what is living in a specific place—though undoubtedly some number of the hawks we count are residents of the Headlands—but those passing through. Birder exemplar Scott Weidensaul points out in *Living on the Wind*: "Bird migration is the one truly unifying natural phenomenon in the world, stitching the continents together in a way that even the great weather systems, which roar out from the poles but fizzle at the equator, fail to do. It is . . . perhaps the most compelling drama in all of natural history."[4] The Golden Gate migratory pathway is but a thread in the overall weave birds are constantly effecting worldwide. Weidensaul says that birds migrate because the earth is tilted, "and as it swings through its orbit each year, first the Northern Hemisphere and then the Southern are pointed toward the sun," and thus we have seasons. Seasons of course bring relative abundance of food and shelter to birds and other species, and the "answer" to why birds migrate is that they are trading off the cost or risks involved with the benefits of getting to seasonally available resources. Yes, there are a few tricky questions here nobody has a good answer for. Why do some birds migrate away from locations that have plentiful resources all year round?

Seventy five to 80 percent of counted hawks and 90 to 95 percent of hawks banded by the GGRO are juvenile, or "hatch-year," and nobody knows exactly why. Maybe once the hawks make it across the Golden Gate, which we anthropomorphically assume is a rather harrowing pass for species that aren't ordinarily aquatic, they figure out how to move along the topographical guidelines of the hills and mountains without resorting to this passage over water. There isn't one southern nesting site for these birds to come from or one northern feeding site for them to go to, however. Many birds sharing the same species designation don't migrate at all. Nobody knows "why why why" about any of this.

"You are wondering, where do the adults go, right?" said Fish when I peppered him with queries. Warning me that he basically had generalities to proffer, he said, "The first answer is that inland migration sites like Goshute, Nevada, and Hawk Mountain, Pennsylvania, tend to record the opposite ratios—adults are seventy-five percent of the count." But the sites on the eastern coasts, like Cape May Point, New Jersey, are like GGRO's, mostly juvenile. This "coastal effect" was identified for songbirds by C. J. Ralph of what is now Point Blue in the 1970s. "The general explanation is that juvenile bird brains have an approximate sense of their migration, a kind of vague structure of where to go and how to get there. This might even have evolutionary value as it allows flexibility and refinement of successful routes over the long term," said Fish. The coast provides an obvious North-South leading line and makes it easier for them to know they're headed in the right direction. Fish added that coastal "lift" on ocean winds has been hypothesized as providing a boost to the birds. "As they get older, birds learn more specific inland routes that favor their behavior and ecology, and these coastal routes tend to move inland." Fish also noted the possibility that up to 50 percent of the juveniles are dying in their first winter of life, after the first fall migration and before the next one; those who survive through this year then continue to survive at a better rate.

Witnessing a single generation perform a time-bound ritual had resonance for me as I was fairly bereft up on the 2015

Hawkwatch. Hawks evolved well before the Pleistocene brought the current climate conditions or something close to them to Hawk Hill, so presumably hawks have been making this journey for at least thirteen thousand years. That's a lot of great-great-grandparents to tally up behind us, a lot of juveniles fledged. The GGRO has a telemetry program and select hawks that are big enough and healthy enough to carry the thing each season are affixed with a tracking device. They are followed around California by volunteers who hop in cars and stay in cheesy motels along the way, until the signal is lost. One favorite GGRO story has a crackerjack team of citizen scientists following a red-tailed hawk assiduously and parking for hours outside a residence in Marin where the hawk had evidently landed. After a while the owner invited the team inside. "Do you know what you have in your backyard?" they asked with awe and excitement. "Yes," the owner replied, reaching into a freezer, pulling out chicken giblets, and popping them into the microwave. This supper was subsequently presented to the hawk waiting for it, as it had been going on four years, outside the back door.

The radio-tracked birds get names and identities, and when the signal is lost, there is a general downbeat, as if just holding on to a transmitted connection with the birds represented a special relationship. In 2012 a broad-winged hawk named Lakota by the telemetry crew was tracked flying between Angel Island and the Marin Headlands, back and forth. The team got ready to follow her to Mexico. Then the transmitter stopped. There was no movement from Lakota, who was recovered on Angel Island, with a puncture wound through her vertebrae and lung. This was presumably the doing of an aggressive competitor hawk. Everybody took this individual mortality of a hawk very hard.

But most of the birds overhead are vastly impersonal. This anonymity is of course what we all ultimately return to, as mortal beings bound for reconfiguration in the carbon cycle, even if it is simply as molecules dispersed through air, water, and soil, as my father is right now. I'm not terribly troubled by ultimate questions like "where do we go after death?" but up on the hill, the sight of this season's pulse of hawks taking their place in a line going back hundreds of millions of years did make me wonder as a child does.

When my father died my mother and brothers immediately started seeing him in hawks. Taking a walk to Gardiners Bay in East Hampton to let the fact of his departure sink in, my brother Eddie reported that he and my mother were dive-bombed by a red-tail, "and that never happens. It was Dad." I didn't say anything. When my brother Jack got on to his second or third father-hawk sighting, I could bear it no longer. "You are not seeing Dad. You are seeing a hawk. You just never paid any attention before—they were always there." My mother laughed. "Mary Ellen, don't deprive us of our Irish superstitions!"

Somebody called a female red-tail. Who are you, hawk? She is the daughter of hawks, I thought. Of course people talk to animals as if they carried the souls of the departed—they get crazed with this sense and I understand that. But hawks of all critters are among the least likely to tolerate extra baggage, psychic or otherwise. Their miracle is their continued existence and their presence on our retinas. Now here, now gone. But then they come again. The same but not the same. Thank you, Heraclitus.

In 2014 I decided to take in hand my inability to identify hawks with anything more than accidental accuracy. I joined the banders. This is the corollary to the Hawkwatch and it is a decidedly more intense commitment, accompanied, as one might expect, by a certain superiority in the tribe. "What, you did the Hawkwatch?" one grizzled veteran of the blinds said to me, as if to say, "You are a wimp." The banding day is considerably longer, considerably colder and hotter, considerably more challenging than the Hawkwatch. We began at 7:30 AM and usually finished up by 7:30 PM. After the day's marching orders we caravanned in cars to the banding blinds, secret lookouts camouflaged by the hillsides.

My first day in a blind I was under the tutelage of John Keane, a very longtime bander and research ecologist with the Forest Service in his day job. Keane is beefy and handsome (a childhood friend of his told me his nickname was Handsome John). The authority conveyed by his brush-cut gray-white hair is supported by jet-black rectangular eyebrows like epaulets. His hands are massive. Keane explained and guided and was mostly patient with me and another apprentice, a woman about my age who is a pilot for Southwest Airlines. Our fourth in the blind was a fairly new bander on his third season, but with extensive experience monitoring condors in Pinnacles National Park. (These birds have a ten-foot wingspan. It takes three people to handle one. Rather than the claws to look out for, with the condor it's the beak.) To Keane's first-in-command of the blind, he was the gentle sergeant who quietly whispered nuanced instructions to guide us step by step through the fairly nerve-racking process of handling and banding a hawk for the first time.

The blind was quiet. The four of us were seated at stools almost shoulder to shoulder in a small shedlike structure barely tall enough to accommodate the height of a tall person. We looked out of horizontal ovals cut in the sides of the structure, mostly straight at the hillside. I was right to think I would start to discern the differences between hawks by assuming this perspective. The red-tails in particular swoop around, as if flirting with the air and the shrubbery, looking keenly interested in a lure bird and then not. They spend a lot of time hanging out in the air, seeming to play with each other, and flying high when they disappear. As I watched them in action up close, their flight habits and behavior started to imprint themselves on my brain and I got a sense of them in situ. Whereas the red-tail seemed to frequently make an appearance from above, Cooper's hawks tended to materialize straight out of the camouflage of the hillside. This much closer to the birds, I found their relative sizes much easier to compare and the smaller falcons were quickly identifiable lightly zipping and zooming past. We waited through long hours of no hawk action in the midafternoon to get to dusk, when harriers were likely to come swooping through. Different hawks use the landscape differently and they also use the light differently.

Into the quiet and meditation of long minutes of staring at birds, sky, and ground, came a whooshing silence, a barked order, and then a clacketing hubbub, as a bow net has been activated and captured a bird, and at least two people have run out to release it as fast as they can. Keane walked back into the blind holding a very large red-tailed hawk. It was a hatch-year female. He looked me in the eye. "I'm going to give you the hawk."

"I have the hawk," I said, once I was holding her with confidence, calmness, and firmness.

"Do you have the hawk?" he reiterated, and I said, "Yes, I have the hawk."

He let go. I was holding the hawk.

The bird was the length of my forearm from elbow to third finger. She was big. I'm holding a hawk, I thought, looking down at her head and beak. She was seemingly indifferent to this turn of events in her day. I looked down farther at her claws, which I maneuvered away from my body and any parts thereof. Her claws were nearly the size of my clenched fists. I held her with one hand and wrote down data points with the other. My internal dialog was about the three Cs: calm, cool, and collected.

Handling the hawk reminded me of giving an infant a bath for the first time and then trying to put a tiny tee shirt on it without dropping the baby, breaking its arm, or freaking out. I summoned the same sense that I had to ground myself and stay centered. The hawk was taking her cues from my vibe.

Even practicing on hawk legs no longer attached to bodies, much less live ones, I had been unable to close the heavier bands with the pliers. My bird was big (so seamlessly I appropriated her!) and she required a big band. I was doing well with all the measurements and weights, but trying to get the band on the hawk, I couldn't do it. "John," I said quietly. Keane looked over and then leaning over, closed the band with his fingers, just like that. Then I was back to the subsequent measurements. I held the hawk in one hand and then that hand slipped ever so slightly. And then I became truly acquainted with the hawk's claws.

The whole thing lasted probably less than thirty seconds, but as in one of those kairos moments of revelation a great deal unfolded therein. This really hurts, it hurts remarkably! I observed, as the hawk squeezed my ring and third fingers and she kept squeezing. This is what she does, right, she squeezes until she breaks a back? How big is the back of a rodent or a small bird and how does the resilience of my fingers compare to their bones? The hawk lives by this action with which she causes death. Frankly, I was enjoying this. I was bonding with this hawk on her terms. Until. "John," I said quietly. "The hawk has my fingers."

Later in the season I sat with Keane in Fort Cronkhite after a day's banding. We went over my apprentice log of skills attained and yet to attain. He pointed out that I had handled buteos, accipiters, and a falcon. He ran his giant index finger—with which he had effortlessly released me from the red-tail's grip—down the little boxes and showed me more I could check off. I felt like a kid getting gold stars. Then his eyebrows came down like a draw-bridge and bifurcated his forehead. "Mary Ellen," he said. "I have a question." This was not going to be a happy question.

"I'm just wondering what's going to happen when we start to get a lot of hawks," he said. There is a fairly predictable surge in the number of hawks in October. A blind can process up to fifty hawks a day during this time. I knew Keane was not questioning my ability to measure, weigh, and otherwise document the hawks, nor was he quibbling over my relative weakness and inability to close the heaviest bands. No, I knew exactly what he was talking about, and it had to do with the capture.

"I've thought about this," I confessed. "The answer is, I don't know. I am afraid I have some ambivalence—like, do I really want to trap a hawk?" Keane's eyebrows stayed horizontal. There had been several occasions when a hawk had come into the bow net I was operating, Keane had given me the order to deploy, and I had not done so in a timely manner. This can be dangerous. If the bow net is sprung while the hawk is moving the metal bar can come down on the bird and injure it. My lack of instinct for capture hadn't imperiled any hawk and Keane noted this. But I stared back at him now. "We want to trap hawks," he said. Like all the great banders, Keane's intensity is matched by his reflexes. They almost scare me, these banders, with how much they want to trap hawks. My reflexes are simply not attached in the same way to a desire to capture a bird. "That's what we're doing. Trapping birds." He placed those hands that hold hawks like baby dolls down on the table. That was it. I was excused.

I decided to head back to the Hawkwatch the next season, and I'm happy to report I took a giant leap forward in my ability to identify birds. "Juvenile red-tail over Elvis!" I found myself shouting recently, to my own great surprise correctly identifying not only the bird but the unofficial landmark over which it was flying—an array of vegetation that by a very large stretch of the imagination looks like the king of rock 'n' roll.

The moment of pain inflicted by my hawk that first day out quickly subsided. I took her outside. My fellow apprentice was as excited as I was. She filmed me with my smartphone and hers. "Look at this bird!" she said. "This is the most beautiful teen-ager I've ever seen! She's the queen!" The bird was completely composed in my hands. Then the thrill of holding the bird was exceeded by the thrill of letting her go. Sighting a hawk can feel like the confirmation of its powers within oneself and holding it is a communion with the same. But the bird wants to fly. The bird has to fly. And so I let go and watched the initial slow flaps of her big wings and then as she soared.

The River Was There

The last story I read to my father was "Big Two-Hearted River." I had been immersed in researching and drafting this book when my father's illness plucked me out of time. I had been discerning the double narrative, the fateful difference between the stories we have been telling ourselves about what we are doing on the landscape and what in fact is going on. Now with the book in my hand under the glow of the bedside table lamp, I was taken aback by the title of the story, as if Hemingway were bouncing my theme back to me over a net. There my father was, listening to the tale, poised between life and death, going over the net. "So here it ends." The thought jumped ahead of me. Of course he was dying.

As my father had commented, Hemingway gets to a language beyond words. He conveys what can't be spoken. As I read the story aloud, I had a sense of all those weighty words, *apocalypse*, *kairos*—even John Steinbeck's *phalanx*, in which the meeting of several minds creates something more. What happened was that Hemingway, my father, and I had a three-way conversation, mostly silent, the upshot of which was "good-bye."

The first two words of the story are "the train," and it is going over a blasted landscape where there is "no town, nothing but the rails and the burned-over country." This is the machine in the blasted garden. The things of man are gone; a fire has destroyed everything. The setting is Michigan woods, but the larger context is World War I. Many a literary critic has parsed this story as the recuperation of Nick Adams after the trauma of war. They see Adams walking like a zombie through the landscape of his destroyed soul, reflected back by evidently reduced nature. It's an okay interpretation, but they put the character's inner life first and I think Hemingway's point is that nature comes first. Hemingway is doing what Joseph Campbell said is the point of myth, to put "the individual in accord with nature."

The main character in the story has things to say but they are intermittent, contextual, and both Nick Adams and the reader have to be especially receptive to hear, as Campbell says, "nature talking." Adams is not all that traumatized. He still has all his bearings. Not needing his map, he "kept on toward the north to hit the river as far upstream as he could go in one day's walking." He measures his ability and ambition scaled to the terrain and the duration of daylight. As I read the story I periodically put my finger on a sentence and looked out the window at the bare trees against the sky. It was the time of year when the foreground darkens first. I stared at the empty message line of the white sky beyond close black trees. My father quizzically waited for the next incident, the "then what" of the story.

Nick watches trout rising up to the surface of the river, "making circles all down the surface of the water, as though it were starting to rain." This beautiful pattern is made by the fish eating insects. Hemingway was a big fan of Cezanne's, and here we have his language doing what Cezanne said art does, making "a harmony parallel to nature." Nick is a presence through which nature is communicated. He is radically alone and not at all. He tunes in to grasshoppers and fish. He thinks about the trees. He is transparent to transcendence, as Campbell would put it. He has broken through, as Ricketts would say.

Hemingway has made a map of Nick's consciousness, taking for its contours the topography and species inhabitants of the woods. The biological inventory is an index or a table of contents outlining both what's out there and what's inside his soul. Hemingway has lined up a science and an art.

Nick eats happily. My father was not exactly enjoying any food by now but he smiled broadly. This is what he would never have again, the simple joy in being alive and tasting it, and his smile was like a salute. In the last line of the story Nick takes the measure of his own future, the time he has left: "There were plenty of days coming when he could fish the swamp." But looking toward those days we see that there will be an end.

I looked at my father and he raised his eyebrows at me, shrugged. He gave a half smile, as if to say, "What can we do?" I thought about what Campbell means when he says the hero's death is the most important part of his journey—it's when we let go of the thing we have clung to, the "I," and in so doing we acquiesce to all that bigger life around and in us. My father was okay with this; he understood what was happening and he was leaning into it. He was losing his life—why do we say it that way? There was more, not less, happening here.

The moment balanced between the past of which my father was every second becoming more a part and the future in which he would not bodily participate. This one story was still unfolding but had essentially already reached and passed its tipping point. Like a village inhabitant rushing into a basement storm cellar, I mentally gathered all the other people I loved and was not losing—children, husband, mother, my siblings and their children, friends. I immediately consoled myself with the evidence of generation. With the fabric of existence and its component parts, the species that make the world. But the black shadow of extinction played in the window curtains. What about the imminent loss of not only thousands of individual plants and animals but also their future kith and kin—the moment still teetered between past and possibility; was not just death at the door but with it the end of birth? In the hero's journey, one life ends to nourish those that will follow. But if this story no longer applies, because we are consuming and not sustaining nature, what is the meaning of individual life?

It was a gift, really, to witness the unfolding of my father's death, renewing as it did my personal sense of meaning in those who remain. These include plants and animals that have not yet reached the end, for which death has yet the possibility of being a generational and not a species-terminating event. If we will but observe them, aggregate their instances and their movements, accommodate and support them, see ourselves as part of and not lords over their world. At the end of *The Hero with a Thousand Faces*, Campbell says historic myths helped humanity figure out the animal and plant worlds, but now man himself is "that alien presence" to be reconfigured. Campbell says that myths have bound us to our tribes and helped consolidate group identities, but now we have to find a way to include all people across all geographies. Confronted with our current extinction event, I wager, Campbell would have said we need to include the larger biotic world in our stories as well. He says the new story will revise the "I" of the old one into a "thou" including all of life. He quotes Nietzsche, "Live as though the day were here." The day is here.

Generation

In spring 2015 I walked with the gang out to Pillar Point to monitor the tide pools. The sea star wasting proceeded apace, but juveniles had been counted in abundance at many sites. As the dawn lightened the world from black to foggy gray, we came upon a stretch of beach littered with worms, about four inches wide and a foot long. They were fringed, it would seem. "Polychaete worms," Rebecca Johnson told us. "They spawn in the high tide. These got beached when the water receded." Several of the worms were sitting in puddles of neon-green goo.

"Those are the gametes," Johnson said. "Some species are cued by the moon and the gametes explode. It kills the worm. Fertilization occurs in the water." Joseph Campbell points to the "inbeingness" of plants in which the seed dies and then generates life. "The plant world is identical in its life sequences with the life of man . . . there's an inward relationship there." Here's an animal with ancestors in the Cambrian that goes about its continuance over millions of years by exploding green stuff in the water. Kind of plantlike when you think about it. Creating more worms! This very raw stuff of life indeed. One of our party snapped a few iNaturalist records.

We moved on to locate and set transects. Close to the breaking waves, about eight people stood stock-still, as if in imitation of twenty or so murres standing on the rocks. I looked and looked, too—and there it was, a spout. Gray whales returning from Mexico with their calves hugged the shore, spouted, rose up, and moved on. Nature felt full and rich, and I silently said hello and good-bye to the whales on their heroic journey.

That night I dreamed I was getting into my minivan with family members as we had a year and a half earlier, the last time my father visited California. In the dream I get in the back behind the driver's seat, which is turned the wrong way. The seats are facing each other and there is no steering wheel. The sense is of a frozen standoff, of not knowing what to do—to get to where the world needs to go. My father sits in the backward driver's seat. He's young and wearing a neon-green suit. We aren't going anywhere, because he is dead. But he is smiling. Be creative, he says, not out loud. Tell the double story.

I woke up from the dream and it was early, dark dissipating moment by moment. The cosmic wheel was turning as it always does. The superreal feeling of my father's presence in the dream had shifted and was now settling down, leaving me bereft, incredulous he could be gone. The data points of his life and times were become lore. Ah! But he was part of the ongoing multiple, the one story that was still unfolding—that day, to be seen and told on Hawk Hill. If not to the stars, I would yet look from the tide pool to the skies. To tell the two stories that are one story. My imminent task was to observe and count birds, my fellow travelers on one heroic journey.

ACKNOWLEDGMENTS

My preoccupation with extinction started at the California Academy of Sciences in 2008. Interviewing most of the scientists there for my book with Susan Middleton, *Evidence of Evolution*, at first I didn't really hear what one after another of them was telling me. I wanted to learn about how life originates, but the scientists wanted to talk about how it is being prematurely terminated. The places they study around the world are changing too fast, they said; their study subjects are disappearing. When one scientist wept talking to me, I thought, gee, he must be having trouble with his wife. When another one teared up, I got a very bad feeling in my stomach. When a third cried, I cried. I won't name them, but I'm still moved to recall the honesty and self-exposure of those three men.

You can't sit with it once you know about extinction. I lit out to write *The Spine of the Continent*, inspired as so many others have been by the vision and lifelong work of Michael Soulé. On the spine I participated in several citizen science projects, and I thought, this could really scale, this could circumvent impediments to saving nature.

Back in California I said, "My next book is about citizen science." Terry Gosliner at the academy said, "Great, we're starting a program, why don't you report on it." Thus I got a ringside seat from which to observe the earnest efforts of a traditional science museum to figure out how best to engage in what is a burgeoning but very young field. I'm grateful to Gosliner, Elizabeth Babcock, Jean Farrington, Alison Young, and Rebecca Johnson for tolerating my constant presence at their elbows.

I wish I had found a way to include Heidi Ballard's top-notch social science research on the efficacy of different citizen science approaches. Tanya Birch and Elizabeth Tyson made important contacts and connections for me. Suzanne Whelan personified the volunteer coordinator who is a combination drill sergeant and mother hen. The insights of many thought leaders have been invaluable and generously shared: these include Healy Hamilton, Lisa Micheli, Greg Newman, Lea Shanley, Rick Bonney, Jennifer Shirk, Jake Weltzin, Geoff McGhee, Adina Merenlender, Fraser Shilling, Gretchen LeBuhn, and Ellie Cohen. (With a few additions, this list comprises a dream team to strategize amplifying the efficacy of citizen science.)

Jon Christensen asked me to report on indigenous burning for *Boom! A Journal of California* and introduced me to the work of Kent Lightfoot, Valentin Lopez, and Doug Bird. I wrote more about the Amah Mutsun for *Bay Nature* magazine with guidance from publisher David Loeb. My understanding of the complexities of the subject would have been far less without insights from Jon and David.

If you have read this book, you know my father died while I was writing it, and that the way he died deeply affected my sense of narrative. Even as he was dying I was thinking, he's doing this in a way that is a gift, an illumination of what life is about upon leaving it. Writing books was for him the be-all, and I'm grateful for his role in creating this one.

Tracy Thorne supported me through my father's death, and even helped me integrate the experience by way of discussing the double narrative. Talking with Brigitte Sandquist about intersecting timescales, it was Brigitte who declared, "It's Chronos and kairos!" My brother Jack Hannibal and friend Alden Mudge gave me solid feedback on early versions of the manuscript. Marilyn Johnson, from whom I first learned how to be an editor at *Esquire* magazine years ago, read two full drafts. Her magic-seeming talent and dedication to the word improved this book and buoyed me emotionally at vulnerable moments. Amanda Pope is another cherished mentor and friend.

A media fellowship from Stanford University's Bill Lane Center for the American West supported the reporting on extinction processes in this book—a foundational contribution for which I'm grateful.

I'm lucky in my agent, Eleanor Jackson, and in the amazing support I have gotten from The Experiment and Workman Publishing. Anne Horowitz's copyedit on the manuscript was extraordinary, as was the meticulous care of my managing editor, Jeanne Tao. My publisher, Matthew Lore, has been a longtime guide and kindred soul. My thanks cannot adequately acknowledge my gifted editor, Nicholas Cizek. With seemingly endless patience and a very light touch, Nick has been by my side looking into the depths, prompting me to clarify until he could see what I was seeing. We have trusted each other, and that has made our work together all the more satisfying.

None of it happens without my husband, Richard. With me he keeps a steady gaze toward a happy, healthy future for our children, Eva and Nick, for generations of *Homo sapiens* yet to come, and for generations upon generations of those species who have come before us, make our life possible, and deserve to fulfill their evolutionary journeys. Gratitude and acknowledgment start with them.

NOTES

Introduction: Change over Time

1 Timothy Morton, *Hyperobjects: Philosophy and Ecology After the End of the World* (Minneapolis: University of Minnesota Press, 2013), 7.

2 Dipesh Chakrabarty, "The Climate of History: Four Theses," in *Critical Inquiry* 35, no. 2 (2009): 201, planetarities.web.unc.edu/files/2015/01/chakrabarty-climate-of-history.pdf

3 Lynn Hunt, *Writing History in the Global Era* (New York: Norton, 2014), 121.

Chapter One: In Which I Freak Out in the Tide Pool

1 Stephen Larsen and Robin Larsen, *A Fire in the Mind: The Life of Joseph Campbell* (New York: Doubleday, 1991), 226.

2 Personal communication with Rich Mooi, curator of invertebrate zoology, California Academy of Sciences.

3 Amanda Stupi, "What We Know—and Don't Know—About the Sea Star Die-Off," KQED Science, March 7, 2014, kqed.org.

4 "Vanishing Fauna," special issue, *Science* 345, no. 6195 (July 25, 2014).

5 National Oceanic Atmospheric Administration Integrated Ecosystem Assessment, "About California Current," noaa.gov/iea/regions/california-current-region/about.html.

6 John Prest, *The Garden of Eden: The Botanic Garden and the Re-Creation of Paradise* (New Haven: Yale University Press, 1981), 9. I love this book. It is fascinating and also beautifully illustrated.

7 I wrote briefly about Linnaeus and Mount Ararat in my previous book, *The Spine of the Continent*. If you happen to have read that book and noticed the repetition, thank you, for one thing. I just think it's worth telling again.

8 Janet Browne drily notes that the "coup de grace for the Ark" was delivered in 1777, in an obscure zoological track by Eberhardt Zimmermann, who pointed out that the pair of lions would have handily eaten the pair of sheep posthaste before turning to the rest of the herbivores on board the Ark. Janet Browne, *The Secular Ark* (New Haven, CT: Yale University Press, 1983), 25.

9 First published as *Journal and Remarks* in 1839.

10 E. O. Wilson, *From So Simple a Beginning: Darwin's Four Great Books* (New York: W. W. Norton, 2006).

11 John Steinbeck, *The Log from the Sea of Cortez* (New York: Penguin Books, 1941), 51.

12 Browne.

13 Charles Darwin, *The Voyage of the* Beagle, 329. Wilson, *From So Simple a Beginning.*

14 Janet Browne, *Charles Darwin: The Power of Place* (New York: Alfred A. Knopf, 2002), 199.

15 The term was coined in 1868 by the English biologist Thomas Huxley.

16 Jared Diamond, "Mr. Wallace's Line," *Discover*, August 1, 1997, discovermagazine.com.

Chapter Two: Moby Ghost

1 New York: St. Martin's, 2016.

2 Barnosky and Hadly, 18–19.

3 Barnosky and Hadly, 19.

4 Marc A. Carrasco, Anthony D. Barnosky, and Russell W. Graham, "Quantifying the Extent of North American Mammal Extinction Relative to the Pre-Anthropogenic Baseline," *PLOS ONE* 4, no. 12 (December 16, 2009).

5 Stephen R. Palumbi and Anthony R. Palumbi, *The Extreme Life of the Sea* (Princeton: Princeton University Press, 2014), 129. The Palumbis describe the otter's air bubbles as "a shimmering silver cloak that warps and twists" as it swims.

6 David Helvarg, *The Golden Shore: California's Love Affair with the Sea* (New York: St. Martin's, 2013), 47.

7 Josie Iselin, *An Ocean Garden: The Secret Life of Seaweed* (New York: Abrams, 2014), 34, 69.

8 Stephen R. Palumbi and Carolyn Sotka, *The Death and Life of Monterey Bay: A Story of Revival* (Washington, DC: Island Press, 2011), 114.

9 Earthwatch is a scientific research institution that routinely uses citizens to help undertake research. Since people pay for the opportunity, Earthwatch doesn't quite fit the definition for citizen science I'm using in this book, which in the broadest sense is free and available to anyone.

10 Morgan Lee, "Marine Revival off San Onofre's Shores," *San Diego Union-Tribune*, March 6, 2014.

11 Michael Nielsen, *Reinventing Discovery: The New Era of Networked Science* (Princeton: Princeton University Press, 2011), 5.

12 Walter Sheldon Tower, *A History of the American Whale Fishery*, Political Economy and Public Law (Philadelphia: Publications of the University of Pennsylvania, 1907), 59.

13 Eric Jay Dolin, *Leviathan: The History of Whaling in America* (New York: W. W. Norton, 2007), 248–49.

14 Matt McGrath, "California Blue Whales Bounce Back to Near Historic Numbers," BBC News, September 5, 2014.

15 Joe Roman et al., "Whales as Marine Ecosystem Engineers," *Frontiers in Ecology and the Environment* 12 (July 2014): 377–85.

16 Craig R. Smith and Amy R. Baco, "Ecology of Whale Falls at the Deep-Sea Floor," *Oceanography and Marine Biology: An Annual Review* 41 (2003): 311–54.

17 Peter Fimrite, "Whale Spotter App to Help Curb Strikes from Ships," *San Francisco Chronicle*, September 29, 2013, sfgate.com.

18 Nadia Drake, "California Shipping Lanes Moved in Attempt to Avoid Killing Whales," *Wired*, May 31, 2013, wired.com.

19 Beach Watch was started by Point Blue in the mid-1980s in response to a major spill and then handed to the sanctuary in the early 1990s.

20 Point Reyes Bird Observatory and David G. Ainley, "The Impacts of the T/V Puerto Rican Oil Spill on Marine Bird and Mammal Populations in the Gulf of the Farallones, 6–19 November, 1984," Stinson Beach, CA: Point Reyes Bird Observatory, March 1985.

21 Historic high common murre die-offs due to starvation have been registered in 2015 and 2016, in turn caused by the unholy triumvirate of the "warm blob" persisting off the West Coast, a powerful El Niño, and climate change patterns, all resulting in the reduction of krill at the bottom of the food chain. See Hannah Hickey, "'Warm Blob' in Pacific Ocean Linked to Weird Weather Across the U.S.," *UW Today*, April 9, 2015, washington.edu.

22 Donald Worster, *Nature's Economy: A History of Ecological Ideas* (New York: Cambridge University Press, 1994), 374–78.

23 T. Fukami et al. "Above- and Below-Ground Impacts of Introduced Predators in Seabird-Dominated Island Ecosystems," *Ecology Letters* 9, no. 12 (December 2006).

24 Amanda Martinez, *Battle at the End of Eden* (Washington, DC: The Atlantic Books, 2012).

25 Bernie R. Tershy et al., "The Importance of Islands for the Protection of Biological and Linguistic Diversity," *BioScience* 6 (June 2015): 1.

26 Barnosky and Hadly, 49.

Chapter Three: The Wild Garden

1 Palumbi and Sotka, 12.

2 Kent G. Lightfoot and Otis Parrish, *California Indians and Their Environment: An Introduction*, California Natural History Guides (Berkeley: University of California Press, 2009), 59. On the matter of bears, they quote the Russian commander Otto von Kotzebue's observations of hunting along the Sacramento River in 1824: "In the night we were much disturbed by bears, which pursued the deer quite close to our tents; and by the clear moonlight we plainly saw a stag spring into the river to escape the bear; the latter, however, jumped after him, and both swam down the stream till they were out of sight." On the subject of quoting Vizcaíno and other early observers, Lightfoot and Parrish caution: "We must be careful about how we interpret these archival documents, since they were written by educated foreign men who often had political and religious motives framing their observations."

3 Juan Crespí, *A Description of Distant Roads: Original Journals of the First Expedition into California, 1769–1770*, trans. and ed. Alan K. Brown (San Diego: San Diego State University Press, 2001), 12.

4 Ibid., 577.

5 Peter S. Alagona, *After the Grizzly: Endangered Species and the Politics of Place in California* (Berkeley: University of California Press, 2013), 20.

6 In April 2016, Guri Dam was the scene of a different sort of nightmare—drought transformed parts of the lake to desert. Venezuelans were subject to electricity rations. Andrew Cawthorne, "Drought-Hit Venezuela Awaits Rain at Crucial Guri Dam," Reuters, April 13, 2016. Historical weather patterns were blamed, but anthropogenic forces have undoubtedly contributed to the situation.

7 Anthony Barnosky, *Dodging Extinction: Power, Food, Money and the Future of Life on Earth* (Berkeley: University of California Press, 2014), 56.

8 While I make an earnest attempt to source my research in primary materials, I probably use Wikipedia to start my search for such several times a day. Thank you, Wikipedia. I do donate to the site, because fair's fair.

9 The phrase *extreme citizen science* was coined by Muki Haklay and Jerome Lewis, codirectors of the Extreme Citizen Science research group at University College London. For a useful blog post discerning the fine points, see Muki Haklay, "Levels of Participation in Citizen Science and Scientific Knowledge Production," *Po Ve Sham*, December 2, 2011, povesham.wordpress.com.

10 Alan Brown, introduction to *A Description of Distant Roads: Original Journals of the First Expedition into California 1769–1770*, ed. Alan K. Brown (San Diego: San Diego State University Press, 2001), 5.

11 Steven W. Hackel, *Junípero Serra: California's Founding Father* (New York: Hill & Wang, 2013), 176.

12 Brown, 3.

13 "History," Amah Matsun Tribal Band site, amahmutsun.org/history.

14 Kimberly Johnston-Dodds, "Early California Laws and Policies Related to California Indians" (Sacramento: California Research Bureau, 2002), 15.

15 Kent G. Lightfoot et al., "Anthropogenic Burning on the Central California Coast in Late Holocene and Early Historical Times: Findings, Implications, and Future Directions," *California Archaeology* 5, no. 2 (December 2013), 371–90.

16 While the Amah Mutsun have thus far been able to ignite just one controlled burn at Pinnacles National Park, other California tribes have been able to do more. Notably Ron Goode, chairman of the North Fork Mono Tribe, has been negotiating successfully with the Forest Service for decades and has lit many fires with the agency.

17 Miller hung out some with Ed Ricketts, who he thought was great, and hit continually on Ricketts' common-law wife, Toni Jackson. Not surprisingly, Miller irritated Steinbeck.

18 Miller is known for the direct-experience, get-out-and-live-your-life credo, but his misreading is basically textual. He is referencing John Keats' poem, "On First Looking into Chapman's Homer," which in turn is referencing another literary text. Keats gave the gift of staggering first sight to Cortez, so at least Miller got that right.

19 M. Kat Anderson, *Tending the Wild: Native American Knowledge and the Management of California's Natural Resources* (Berkeley: University of California Press, 2013), 128.

20 Joseph Campbell with Bill Moyers, *The Power of Myth* (New York: Anchor Books, 1988), 65.

21 Ibid., 47, 45.

Chapter Four: Dream Machine

1 Aldo Leopold, *A Sand County Almanac* (New York: Oxford University Press, 1989), 204.

2 Rick Bonney at Cornell is credited with using the term *citizen science* beginning in the mid-1990s. In 1995, Alan Irwin published *Citizen Science: A Study of People, Expertise and Sustainable Development* (London: Routledge). See also Muki Haklay's chapter, "Citizen Science and Volunteered Geographic Information: Overview and Typology of Participation," in *Crowdsourcing Geographic Knowledge*, Daniel Sui, Sarah Elwood, Michael Goodchild, eds. (New York: Springer, 2013), 105–22.

3 The Second and Third US Open Government National Action Plans and the US National Civil Earth Observations Strategic Plan.

4 Leo Marx, *The Machine in the Garden: Technology and the Pastoral Ideal in America* (New York: Oxford University Press, 1964), 118.

5 Ibid., 134.

6 Ibid., 130.

7 Henry Nash Smith, *Virgin Land* (Cambridge, MA: Harvard University Press, 1978), 8.

8 Accessible discussions of the subject are to be found in the works of Jared Diamond and Charles C. Mann.

9 Scott Weidensaul, *Living on the Wind: Across the Hemisphere with Migratory Birds* (New York: North Point Press, 1999), 78.

10 Stephen E. Ambrose, *Undaunted Courage: Meriwether Lewis, Thomas Jefferson, and the Opening of the American West* (New York: Touchstone, 1996), 70–71. Bernard DeVoto, *The Course of Empire* (Boston: Houghton Mifflin, 1952), 346.

11 Quoted in Ambrose, 76.

12 Mark Stein, *How the States Got Their Shapes* (New York: HarperCollins, 2008), 35.

13 Mary Hill, *Geology of the Sierra Nevada*, rev. ed., California Natural History Guides (Berkeley: University of California Press, 2006), 213.

14 Ibid., 219.

15 Ibid., 214.

16 Michael Smith, *Pacific Visions: California Scientists and the Environment, 1850–1915* (New Haven: Yale University Press, 1987), 76.

17 Thomas E. Sheridan and Nathan F. Sayre, "A Brief History of People and Policy in the West," in *Stitching the West Back Together: Conservation of Working Landscapes*, Summits: Environmental Science, Law, and Policy, eds. Susan Charnley, Thomas E. Sheridan, and Gary P. Nabhan (Chicago: University of Chicago Press, 2014), 3–9.

Chapter Five: Green Thumb in a Dark Eden

1 Saurav Dhakal, "In Pursuit of Stories," *Nepali Times*, March 27, 2015, nepalitimes.com.

2 Rebecca Moore's middle name is Tilden.

3 "Samuel J. Tilden Papers 1794–1886," Kit Messick, comp. New York Public Library Manuscripts and Archives Division, March 2010, http://archives.nypl.org/mss/2993.

4 Kay Mills, *Changing Channels: The Civil Rights Case That Transformed Television* (Jackson: University Press of Mississippi, 2004).

5 Rebecca Moore, "The Democratization of Geospatial Information: Ideas for the Google Earth Project" (talk, Google, Mountainview, CA, May 31, 2005).

6 Eric Sheppard, "GIS and Society: Towards a Research Agenda," *Cartography and Geographic Information Systems* 22, no. 1 (1995), 5–16.

Chapter Six: Into the Woods

1 William Gibbons, "The Redwood in the Oakland Hills," *Erythea: A Journal of Botany, West American and General* 1, no. 8 (August 1, 1893), 166.

2 James A. Johnstone and Todd E. Dawson, "Climatic Context and Ecological Implications of Summer Fog Decline in the Coast Redwood Region," *Proceedings of the National Academy of Sciences* 107, no. 10 (February 2010).

3 Bereket Lebassi et al., "Observed 1970–2005 Cooling of Summer Daytime Temperatures in Coastal California," *Journal of Climate* 22 (November 2008), 3558–73.

4 Susan R. Schrepfer, *The Fight to Save the Redwoods* (Madison: University of Wisconsin Press, 1983), 3.

5 Ibid., 5.

6 As Joseph Campbell puts it in *The Hero with a Thousand Faces*, 223, "God assumes the life of man and man releases the God within himself at the mid-point of the cross-arms of the same 'coincidence of opposites,' the same sun door through which God descends and Man ascends—each as the others' food." Kind of a trophic gloss on the whole thing as well.

7 Schrepfer, 45.

8 In fact the words don't quite appear that way in Goethe. Thomas Carlyle essentially paraphrases Faust in *Sartor Resartus* and the resulting encomium has stuck. Thomas Carlyle, *Sartor Resartus* (Boston: Athenaeum Press, 1876), 48.

9 Schrepfer, 39.

10 Ibid., 40.

Chapter Seven: We All Want to Change the World

1 The primary source for information regarding the inception of the academy is Theodore Henry Hittell's *The California Academy of Science: 1853-1906*, the manuscript of which was among the treasures rescued by Alice Eastwood from the academy after the 1906 earthquake and right before the fire destroyed the building. Its current format, published by

the academy in 1997, was prepared by Al Leviton and Michele Aldrich, who spent ten years verifying Hittell and annotating his document.

2 Theodore Henry Hittell, *The California Academy of Sciences: A Narrative History* (San Francisco: California Academy of Sciences, 1997), 489.

3 Barbara Ertter, "People, Plants, and Politics: The Development of Institution-Based Botany in California 1853-1906," Cultures and Institutions of Natural History (San Francisco: California Academy of Sciences, 2000). Ertter is the semiretired curator of western North American flora at the University of California, Berkeley.

4 This species name is currently "unaccepted" by the World Register of Marine Species—I'm trying to figure out what that means!

5 Al Leviton, emeritus curator at today's academy as well as director of its scientific publications, told me that Kellogg was "a botanist of the first order for his time and he described many new species."

6 H. H. Behr, *Synopsis of the Genera of Vascular Plants in the Vicinity of San Francisco, with an Attempt to Arrange Them According to Evolutionary Principles* (San Francisco: Payot, Upham, & Co., 1884), 8. I uploaded this text and printed out a PDF of some of it via the Biodiversity Heritage Library, part of the Internet Archive, which makes millions of pages of historical text available to all of us: SO COOL. archive.org/details/biodiversity.

7 Nabokov was himself a citizen scientist. With no formal training, he yet made significant contributions to lepidopterology that are even today shedding new light on earth history. He ran the lepidoptera collection at Harvard's Museum of Comparative Zoology while teaching literature at Wellesley in the 1940s.

8 Elizabeth B. Keeney, *The Botanizers: Amateur Scientists in Nineteenth-Century America* (Chapel Hill: University of North Carolina Press, 1992), 102-9.

9 Ertter, 216, quoting William H. Goetzmann, *Exploration and Empire: The Explorer and the Scientist in the Winning of the American West* (New York: Knopf, 1966), 355.

10 Ibid., 207, 209.

11 Browne, *Charles Darwin: The Power of Place*, 38.

12 Hittell, 29.

13 Browne, *Charles Darwin: The Power of Place*, 50-52.

14 Hittell, 146.

15 1905-06 Galápagos Expedition papers, California Academy of Sciences Archives, San Francisco, CA.

16 Field Notebook, Joseph Richard Slevin papers, Special Collections, California Academy of Sciences Library, San Francisco, CA.

17 John P. Dumbacher and Barbara West, "Collecting Galápagos and the Pacific: How Rollo Howard Beck Shaped Our Understanding of Evolution," *Proceedings of the California Academy of Sciences* 61, no. 13 (September 15, 2010): 211-43. Dumbacher is curator of birds and mammals at the academy, and citizen birders who would like to take his class in

bird identification, be advised that entry requires you identify five hundred species by sight at the get-go. West has a special place in Galápagos history, since she is among those making sure the islands have one. She has almost single-handedly compiled, organized, and sent on their way to digitization the archives of the 1905-06 expedition in the academy's library. West has worked on behalf of protecting the Galápagos for almost thirty years.

18 The Webster-Harris Expedition is named for Frank Blake Webster, a commercial purveyor of animals to zoos and museums, who didn't take the trip, and Charles Harris, a taxidermist who sufficiently recovered from a bout of yellow fever to take leadership of the expedition when the captain died of it.

19 Miriam Rothschild, *Dear Lord Rothschild: Birds, Butterflies and History* (Glenside, PA: Balaban, 1983).

20 Matt James interview with Roger Wolfe, April 5, 2006, montereyseabirds .com/RolloBeckBio.htm.

21 Field Notebook, Joseph Richard Slevin papers.

22 Ibid.

23 Ibid.

24 Dumbacher and West, 13.

25 Joseph Grinnell, "Geography and Evolution," *Ecology* 5, no. 3 (July 1924): 225-29.

26 Joseph Grinnell, "Uses of a Research Museum," in *Joseph Grinnell's Philosophy of Nature: Selected Writings of a Western Naturalist* (Berkeley: University of California Press, 1943), 35.

27 Craig Moritz et al., "Impact of a Century of Climate Change on Small-Mammal Communities in Yosemite National Park, USA," *Science* 322 (October 2008), 261-64.

28 Dumbacher and West, 227.

29 Ernst Mayr, *The Growth of Biological Thought: Diversity, Evolution, and Inheritance* (Cambridge, MA: Harvard University Press, 1982), 275.

30 From Darwin's autobiography, published in 1958 by his granddaughter Nora Barlow: "I saw two rare beetles, and seized one in each hand; then I saw a third and new kind, which I could not bear to lose, so that I popped the one which I held in my right hand into my mouth. Alas! it ejected some intensely acrid fluid, which burnt my tongue so that I was forced to spit the beetle out, which was lost, as was the third one."

31 The other two graduate students were Nate Agrin and Jessica Klein.

32 Scott R. Loarie et al., "Climate Change and the Future of California's Endemic Flora," PLOS ONE 3, no. 6 (June 25, 2008).

33 It also led to funding of a broader project to assess California's biotic adaptation to climate change, discussed later on in this book.

34 S. L. Pimm et al., "The Biodiversity of Species and Their Rates of Extinction, Distribution, and Protection," *Science* 344, no. 6187 (May 30, 2014).

35 Roger Yu, "Plains All American Pipeline Indicted After California Oil Spill," *USA Today*, May 18, 2016, usatoday.com.

36 James R. Griesemer and Elihu M. Gerson, "Collaboration in the Museum of Vertebrate Zoology," *Journal of the History of Biology* 26, no. 2 (summer 1993), 185–203.

37 Barbara R. Stein, *On Her Own Terms: Annie Montague Alexander and the Rise of Science in the American West* (Berkeley: University of California Press, 2001), 29.

38 Ibid., 49.

39 Ibid., 47.

40 Ibid., 249.

41 Roberta Millstein, "Environmental Ethics," in *The Philosophy of Biology: A Companion for Educators*, ed. Kostas Kampourakis (New York: Springer, 2013), 723–41.

42 Aldo Leopold, *A Sand County Almanac* (New York: Oxford University Press, 1949), 105–7.

43 Herman Melville, *Moby-Dick*, Norton Critical Edition (New York: W. W. Norton, 2002), 139–40.

44 Campbell and Moyers, *The Power of Myth*, xiv.

45 Richard Slotkin, *Regeneration Through Violence: The Mythology of the American Frontier 1600–1860* (Norman: University of Oklahoma Press, 1973), 22.

Chapter Eight: First There Is a Mountain

1 Tom Killion and Gary Snyder, *Tamalpais Walking: Poetry, History, and Prints* (Berkeley: Heyday Books, 2009), 11.

2 Ibid., 44.

3 Campbell, *The Hero's Journey*, 13.

4 The town of Kentfield in Marin County was named for William's father, the original moneybags who in the late 1800s brought a livestock fortune west from Chicago. Killion and Snyder, p. 81.

5 Ibid., 4.

6 Ibid., 19.

7 Robert Hass, foreword, *The Place That Inhabits Us: Poems of the San Francisco Bay Watershed*, ed. Sixteen Rivers Press (San Francisco: Sixteen Rivers Press, 2010), ix–xiii.

8 Thomas F. Daniel, "One Hundred and Fifty Years of Botany at the California Academy of Sciences (1853–2003)," *Proceedings of the California Academy of Sciences* 59, no. 7 (May 16, 2008): 215–305.

9 "The California Academy of Sciences," *Science* 23, no.595 (May 25, 1906): 823–26.

10 Doris Sloan, *Geology of the San Francisco Bay Region*, California Natural History Guides (Berkeley: University of California Press, 2006).

11 Kent Lightfoot and Otis Parrish, *California Indians and their Environment* (Berkeley: University of California Press, 2009), 52.

12 Killion and Snyder, 43.

13 Alice Eastwood to Joseph Grinnell, October 30, 1917, California Academy of Sciences Archives, Eastwood correspondence, box 49.

14 Michael Ghiselin, "The Individual in the Darwinian Revolution," *New Literary History* 3, no. 1 (autumn 1971):113-34.

15 Robert Pogue Harrison, *Gardens: An Essay on the Human Condition* (Chicago: University of Chicago Press, 2008), 60, 74.

16 Daniel, 268-69.

17 Naomi Klein, "A Radical Vatican?" *The New Yorker*, July 10, 2015, newyorker.com.

18 Paul R. Ehrlich and Peter Raven, "Butterflies and Plants: A Study in Coevolution," *Evolution* 18 (1964): 588.

19 Ibid., 606.

20 Paul R. Ehrlich and Ilkka Hanski, eds., *On the Wings of Checkerspots: A Model System for Population Biology* (New York: Oxford University Press, 2004), 588.

21 Limberbush is actually less flexible in response to changing seasons than brittlebush. It does not leaf out until heavy monsoon rains begin. "It's like the oak of the desert," Weltzin told me.

22 As far as we know, indigenous people did not burn in the desert.

Chapter Nine: Innocence and Experience

1 It's possible this quote is apocryphal—exactly when Campbell stated this is difficult to confirm—but it's widely used and always attributed to Campbell.

2 "The Hero's Adventure: A Tribute to Joseph Campbell by Michael Toms," New Dimensions Radio Network News 15, January–February 1988.

3 Jamake Highwater, "A Conversation with Joseph Campbell," *Quadrant* 18, no. 1 (spring 1985).

4 Fraser Boa, *The Way of the Myth: Talking with Joseph Campbell* (Boston and London: Shambhala, 1994), 5.

5 John McCosker, "The View from the Great Tidepool," *Pacific Discovery* 40, no. 4 (October/December 1987): 35.

6 Ed Ricketts, "Vagabonding Through Dixie," *Travel* 45, no. 2 (June 1925).

7 John Muir, *A Thousand-Mile Walk to the Gulf* (New York: Houghton Mifflin Company, 1916), 75.

8 Smith, *Pacific Visions*, 224.

9 Ibid., 94.

10 John Muir, "Yosemite Glaciers," *New York Tribune*, December 5, 1971.

11 Quoted in the introduction to *Breaking Through: Essays, Journals, and Travelogues of Edward F. Ricketts*, ed. Katharine A. Rodger (Berkeley: University of California Press, 2006), 4.

12 Gregg Mitman, *The State of Nature: Ecology, Community, and American Social Thought 1900-1950* (Chicago: University of Chicago Press, 1992).

13 Jackson J. Benson, *The True Adventures of John Steinbeck, Writer* (New York: Penguin Books, 1984), 192.

14 Ibid., 192.

15 Rodger, 7.

16 Benson, 194.

17 Rodger, 82–83.

18 Ibid., 11.

19 From 1996 to her early death in 2004, Dr. Molly Ahlgren initiated intertidal surveys with schoolchildren in Sitka that have been incorporated into the overall data. An emergency medical technician, Ahlgren died in a boating accident while responding to a call with the Sitka Fire Department.

20 Jack Calvin, "'Nakwasina' Goes North," *National Geographic* 64, no. 1 (July 1933), 1.

21 Benson, 167.

22 Jay Parini, *John Steinbeck: A Biography* (New York: Henry Holt, 1995), 101.

23 Benson, 184.

24 Susan Shillinglaw, *Carol and John Steinbeck: Portrait of a Marriage,* Western Literature Series (Reno: University of Nevada Press, 2013), 81.

25 Quoted in Benson, 241.

26 Shillinglaw, 130.

27 Ibid., 95.

28 Ibid., 91.

29 Ibid., 92. Shillinglaw is quoting Gustaf Lannestock's recollection of the evening.

30 Quoted in Michael J. Lannoo, *Leopold's Shack and Ricketts's Lab* (Berkeley: University of California Press, 2010), 71.

31 Quoted in Larsen and Larsen, 167.

32 Ibid., 176.

33 Ibid., 38–39.

34 His advisor, Raymond Weaver, is famous for finding the manuscript of *Billy Budd* in a drawer while he was undertaking Herman Melville's first biography.

35 Boas is considered the progenitor of cultural relativism, or the postulate that one culture cannot be considered better or less good than another, and that all cultures create their own norms. Boas eventually established a graduate program in anthropology at Columbia, and his first graduate student was Alfred Kroeber, who founded UC Berkeley's department in the discipline. Among Boas' students were Margaret Mead and Zora Neale Hurston.

36 Quoted in Larsen and Larsen, 89.

37 Dark Mountain Project, dark-mountain.net. For more on the Dark Mountain Project, see Daniel Smith, "It's the End of the World as We Know It . . . and He Feels Fine," *The New York Times Magazine*, April 17, 2014, nytimes.com.

38 Quoted in Parini, 136.

39 "The Hero's Adventure," New Dimensions Radio Network News.

40 From Campbell's "Grampus Diary," quoted in Larsen and Larsen, 186. Campbell's papers are collected in the Opus Archives, and while the Grampus Diary has in the past been made available to researchers, it

is currently withheld by the estate. There is a question of whether the material is fiction or nonfiction. Shillinglaw said it is clear Campbell was writing fiction based on the material, but that the fiction is a separate manuscript.

41 Shillinglaw, 113.

42 Eric Enno Tamm, *Beyond the Outer Shores: The Untold Odyssey of Ed Ricketts, the Pioneering Ecologist Who Inspired John Steinbeck and Joseph Campbell* (New York: Da Capo Press, 2005), 15.

43 Quoted in Larsen and Larsen, 203.

44 Tamm, 184–86.

45 Shillinglaw, 115.

46 Ibid., 160.

47 Joseph Campbell, *The Hero's Journey: Joseph Campbell on His Life and Work*, ed. Phil Coisineau (San Francisco: Harper & Row, 1990).

48 Larsen and Larsen, 255.

49 Susan F. Beegel, Susan Shillinglaw, Wesley N. Tiffney Jr., *Steinbeck and the Environment: Interdisciplinary Approaches* (Tuscaloosa: University of Alabama Press), 10.

50 Edmund Wilson, "The Californians: Storm and Steinbeck," in Joseph R. McElraith, Jesse S. Crisler, and Susan Shillinglaw, eds., *John Steinbeck: The Contemporary Reviews* (New York: Cambridge University Press, 1996), 168.

51 McCosker, 34–41.

52 Quoted in Susan F. Beegel's foreword to *Breaking Through*, Rodger ed.

53 Ironically, Covici eventually became Joseph Campbell's editor as well, and Steinbeck and Campbell were destined to meet again at his funeral, where the former rivals conversed with warmth.

54 Steinbeck, *The Log from the Sea of Cortez*, xxvii.

55 Quoted by Richard Astro in the introduction to *The Log from the Sea of Cortez*, xiii, xx.

56 Steinbeck, *The Log from the Sea of Cortez*.

57 Wesley N. Tiffney Jr., "Introduction," in *Steinbeck and the Environment: Interdisciplinary Approaches*, ed. Susan F. Beegel, Susan Shillinglaw, and Wesley N. Tiffney Jr. (Tuscaloosa: University of Alabama Press, 1997), 12–13.

58 Quoted in Tamm, 192.

59 Joel W. Hedgpeth, "John Steinbeck: Late-Blooming Environmentalist," in Beegel, Shillinglaw, and Tiffney, *Steinbeck and the Environment*, 301.

60 William F. Gilly, "John Steinbeck, Ed Ricketts & Holistic Biology: A New Voyage to the Sea of Cortez," *Phi Beta Kappa: The Key Reporter* (summer 2006): 10–11.

61 William Gilly, "Searching for the Spirits of the Sea of Cortez," *Steinbeck Studies* 15, no. 2 (fall 2004): 5–14.

62 Rafe Sagarin and Aníbel Pauchard, *Observation and Ecology: Broadening the Scope of Science to Understand a Complex World* (Washington, DC: Island Press, 2012).

63 R. Sagarin et al., "Between Control and Complexity: Opportunities and Challenges for Marine Mesocosms," *Frontiers in Ecology and the Environment.* God bless them, the titles of even outlier biologists are a slog.

64 Rafe Sagarin, "On Observation: Notes from a Field Course on the California Coast," *Boom: A Journal of California* 4, no. 3 (Fall 2014): 113–21.

Chapter Ten: Bee, I'm Expecting You

1 Andy Isaacson, "Three Generations of Citizen Science: The Incubator," *Audubon*, November–December 2014, audubon.org.

2 William Newmark, "A Land-Bridge Island Perspective on Mammalian Extinctions in Western North American Parks," *Nature* 325 (January 27 1987), 430–32.

3 Robbins started the Breeding Bird Survey in the 1960s, and it was quickly taken over by Danny Bystrak. Droege counts Bystrak a mentor as well.

4 Jeff Hull, "Three Generations of Citizen Science: The Pioneer," *Audubon*, November–December 2014, audubon.org.

5 Jeff Selleck, "National Parks and Biodiversity Discovery," National Park Service, *Park Science* 31, no. 1 (Special Issue 2014).

6 Beth Slatkin, "Spreading the Buzz About Native Bees," *Bay Nature*, June 19, 2014, baynature.org.

7 Sam Droege et al., "Pollinators in Peril? A Multipark Approach to Evaluating Bee Communities in Habitats Vulnerable to Effects from Climate Change," *Park Science* 31, no. 1 (2014).

8 Matt Williams, "Will Earth Survive When the Sun Becomes a Red Giant?" *Universe Today*, May 9, 2016, universetoday.com.

9 Harvey Locke, "Nature Needs Half: A Necessary and Hopeful New Agenda for Protected Areas," *Parks* 19, no.1 (March 2013): 9.

Chapter Eleven: Eyewitness

1 They don't have "great horns," but they are pretty great.

2 I was at a workshop on climate change filled with scientists the day Schneider's sudden death was announced. The room hushed. Most people bent their heads and at least one straight-ahead face turned red and wet. It felt like more than the death of a person. The moment challenged and intensified the purpose in the room.

3 Camille Parmesan *Science Watch* Newsletter Interview, March 2010, archive.sciencewatch.com/inter/aut/2010/10-mar/10marParm.

4 Scott Weidensaul, *Living on the Wind: Across the Hemisphere with Migratory Birds* (New York: North Point Press, 1999), x.

SELECT BIBLIOGRAPHY

Alagona, Peter S. *After the Grizzly: Endangered Species and the Politics of Place in California.* Berkeley: University of California Press, 2013.

Ambrose, Stephen E. *Undaunted Courage: Meriwether Lewis, Thomas Jefferson, and the Opening of the American West.* New York: Simon & Schuster, 1996.

Anderson, M. Kat. *Tending the Wild: Native American Knowledge and the Management of California's Natural Resources.* Berkeley: University of California Press, 2005.

Asma, Stephen T. *Stuffed Animals and Pickled Heads: The Culture and Evolution of Natural History Museums.* Oxford: Oxford University Press, 2001.

Astro, Richard. *John Steinbeck and Edward F. Ricketts: The Shaping of a Novelist.* Minneapolis: University of Minnesota Press, 1973.

Barnes, R. S. K., and K. H. Mann, eds. *Fundamentals of Aquatic Ecology.* Oxford: Blackwell Science, 1980.

Barnosky, Anthony D. *Dodging Extinction: Power, Food, Money, and the Future of Life on Earth.* Berkeley: University of California Press, 2014.

Barnosky, Anthony D., and Elizabeth A. Hadly. *End Game: Tipping Point for Planet Earth.* New York: St. Martin's, 2016.

Beegel, Susan F., Susan Shillinglaw, and Wesley N. Tiffney Jr., eds. *Steinbeck and the Environment: Interdisciplinary Approaches.* Tuscaloosa: University of Alabama Press, 1977.

Benson, Jackson J. *The True Adventures of John Steinbeck, Writer.* New York: Penguin, 1984.

Blunt, Wilfred. *Linnaeus: The Compleat Naturalist.* Princeton: Princeton University Press, 2001.

Boggs, Carol L., Ward B. Watt, and Paul R. Ehrlich, eds. *Butterflies: Ecology and Evolution Taking Flight.* Chicago: University of Chicago Press, 2003.

Browne, Janet. *The Secular Ark: Studies in the History of Biogeography.* New Haven: Yale University Press, 1983.

———. *Charles Darwin: The Power of Place.* New York: Alfred A. Knopf, 2002.

———. *Charles Darwin: Voyaging.* New York: Alfred A. Knopf, 1995.

Brown, Joseph Epes. *The Spiritual Legacy of the American Indian with Letters While Living with Black Elk.* Bloomington: World Wisdom, 2007.

Burns, Loree Griffin, and Ellen Harasimowicz. *Citizen Scientists: Be a Part of Scientific Discovery from Your Own Backyard.* New York: Henry Holt, 2012.

California Coastal Commission. *California Coastal Access Guide,* 7th ed. Berkeley: University of California Press, 2014.

Campbell, Joseph. *The Hero with a Thousand Faces*. New York: Pantheon, 1949.

———. *The Hero's Journey: Joseph Campbell on His Life and Work*. Novato: New World Library, 1990.

———. *Myths to Live By*. New York: Penguin, 1972.

Campbell, Joseph, with Bill Moyers. *The Power of Myth*. New York: Anchor, 1991.

Canfield, Michael R., ed. *Field Notes on Science and Nature*. Cambridge: Harvard University Press, 2011.

Cannon, Susan Faye. *Science in Culture: The Early Victorian Period*. New York: Neale Watson Academic Publications, 1978.

Carle, David. *Introduction to Earth, Soil, and Land in California*. California Natural History Guides. Berkeley: University of California Press, 2010.

———. *Introduction to Fire in California*. California Natural History Guides. Berkeley: University of California Press, 2008.

Carlton, James T., ed. *The Light and Smith Manual: Intertidal Invertebrates from Central California to Oregon*, 4th ed. Berkeley: University of California Press, 2007.

Clark, William S., and Brian K. Wheeler. *Hawks of North America*, 2nd ed. New York: Houghton Mifflin, 2001.

Corburn, Jason. *Street Science: Community Knowledge and Environmental Health Justice*. Urban and Industrial Environments. Cambridge, MA: MIT Press, 2005.

Crespí, Juan. *A Description of Distant Roads: Original Journals of the First Expedition into California, 1769–1770*. Alan K. Brown, ed. San Diego: San Diego State University Press, 2001.

Crossley, Richard, Jerry Ligouri, and Brian Sullivan. *The Crossley ID Guide: Raptors*. Princeton: Princeton University Press, 2013.

Dakin, Susanna Bryant. *The Perennial Adventure: A Tribute to Alice Eastwood, 1859–1953*. San Francisco: California Academy of Sciences, 1953.

Davis, Matthew, and Michael Farrell Scott. *Opening the Mountain: Circumambulating Mount Tamalpais, a Ritual Walk*. Emeryville: Shoemaker & Hoard, 2006.

De Nevers, Greg, Deborah Stanger Edelman, and Adina Merenlender. *The California Naturalist Handbook*. Berkeley: University of California Press, 2013.

Dickinson, Janis L., and Rick Bonney. *Citizen Science: Public Participation in Environmental Research*. Ithaca: Cornell University, 2012.

Dumbacher, John P., and Barbara West. "Collecting Galápagos and the Pacific: How Rollo Howard Beck Shaped Our Understanding of Evolution." *Proceedings of the California Academy of Sciences* 61, no. 13 (September 15, 2010): 211–43.

Ehrlich, Paul R., and Anne Ehrlich. *Extinction: The Causes and Consequences of the Disappearance of Species*. New York: Random House, 1981.

Elton, Charles. *Animal Ecology.* Chicago: University of Chicago Press, 2001.

———. *The Ecology of Invasions by Animals and Plants.* Chicago: University of Chicago Press, 1958.

Ellis, Richard. *Men and Whales.* New York: Knopf, 1991.

Farmer, Jared. *Trees in Paradise: A California History.* New York: W. W. Norton, 2013.

Faunt, Claudia C., ed. *Groundwater Availability of the Central Valley Aquifer, California.* US Geological Survey Professional Paper 1776, 2009.

Ghiselin, Michael T. "Darwin: A Reader's Guide." Occasional Papers of the California Academy of Sciences no. 155. San Francisco: California Academy of Sciences, 2009.

Greig-Smith, P. *Quantitative Plant Ecology,* 3rd ed. Studies in Ecology 9. Berkeley, University of California Press, 1983.

Gutiérrez, Ramón A., and Richard J. Orsi, eds. *Contested Eden: California Before the Gold Rush.* Berkeley: University of California Press, 1998.

Harrison, Robert Pogue. *Gardens: An Essay on the Human Condition.* Chicago: University of Chicago Press, 2008.

Hughes, Patrick. *A Century of Weather Service: A History of the Birth and Growth of the National Weather Service, 1870–1970.* New York: Gordon and Breach, 1970.

Hunt, Lynn. *Writing History in the Global Era.* New York: W. W. Norton, 2014.

Jefferson, Thomas. *Notes on the State of Virginia.* New York: Penguin, 1999.

Kerlinger, Paul, and Pat Archer. *How Birds Migrate,* 2nd ed. Mechanicsburg: Stackpole, 2008.

Kertess, Klaus, Susan Harris, and Gregg Bordowitz. *Toxic Beauty: The Art of Frank Moore.* New York: Grey Art Gallery, New York University, 2012.

Killion, Tom, and Gary Snyder. *Tamalpais Walking: Poetry, History, and Prints.* Berkeley: Heyday, 2009.

La Pérouse, Jean François de. *Life in a California Mission: Monterey in 1786.* Berkeley: Heyday, 1989.

Lack, David. *The Life of the Robin.* London: Witherby, 1943.

Larsen, Stephen, and Robin Larsen. *Joseph Campbell: A Fire in the Mind.* Rochester: Inner Traditions, 2002.

Leopold, Aldo. *A Sand County Almanac.* New York: Oxford University Press, 1949.

Lewis, Meriwether, and William Clark. *The Journals of Lewis and Clark,* rev. ed. Bernard DeVoto, ed. New York: Mariner, 1997.

Lightfoot, Kent G. *Indians, Missionaries, and Merchants: The Legacy of Colonial Encounters on the California Frontiers.* Berkeley: University of California Press, 2005.

Lightfoot, Kent G., and Otis Parrish. *California Indians and their Environment: An Introduction.* California Natural History Guides. Berkeley: University of California Press, 2009.

Lisca, Peter. *John Steinbeck: Nature and Myth*. New York: Thomas Y. Crowell, 1978.

Losos, Jonathan B., and Robert E. Ricklefs, eds. *The Theory of Island Biogeography Revisited*. Princeton: University of Princeton Press, 2010.

Lowen, Rebecca S. *Creating the Cold War University: The Transformation of Stanford*. Berkeley: University of California Press, 1997.

MacArthur, Robert H. *Geographical Ecology: Patterns in the Distribution of Species*. Princeton: Princeton University Press, 1972.

MacArthur, Robert H., and Edward O. Wilson. *The Theory of Island Biogeography*. Princeton: Princeton University Press, 1967.

Martin, Edwin T. *Thomas Jefferson: Scientist*. New York: Henry Schuman, 1952.

Marx, Leo. *The Machine in the Garden: Technology and the Pastoral Ideal in America*. New York: Oxford University Press, 1964.

Milliken, Randall. *A Time of Little Choice: The Disintegration of Tribal Culture in the San Francisco Bay Area 1769–1810*. Menlo Park, CA: Ballena Press, 1995.

Mills, Kay. *Changing Channels: The Civil Rights Case That Transformed Television*. Jackson: University Press of Mississippi, 2004.

Minnich, Richard A. *California's Fading Wildflowers: Lost Legacy and Biological Invasions*. Berkeley: University of California Press, 2008.

Mount, Jeffrey F. *California Rivers and Streams: The Conflict Between Fluvial Process and Land Use*. Berkeley: University of California Press, 1995.

Nice, Margaret Morse. *The Watcher at the Nest*. New York: Macmillan, 1939.

Nielsen, Michael. *Reinventing Discovery: The New Era of Networked Science*. Princeton: Princeton University Press, 2012.

Noss, Reed F., ed. *The Redwood Forest: History, Ecology, and Conservation of the Coast Redwoods*. Washington, DC: Island Press, 2000.

Oelschlaeger, Max. *The Idea of Wilderness: From Prehistory to the Age of Ecology*. New Haven: Yale University Press, 1991.

Ornduff, Robert, Phyllis M. Faber, and Todd Keeler-Wolf. *Introduction to California Plant Life*, rev. ed. California Natural History Guides. Berkeley: University of California Press, 2003.

Palumbi, Stephen R., and Carolyn Sotka. *The Death and Life of Monterey Bay: A Story of Revival*. Washington, DC: Island Press, 2011.

Parini, Jay. *John Steinbeck: A Biography*. New York: Henry Holt, 1996.

Phillips, Julie, et al. *Safe Passage for Coyote Valley: A Wildlife Linkage for the Highway 101 Corridor*. Cupertino: Kirsch Center for Environmental Studies, De Anza College, 2012.

Prest, John. *The Garden of Eden: The Botanic Garden and the Re-Creation of Paradise*. New Haven: Yale University Press, 1988.

Pyne, Stephen J. *World Fire: The Culture of Fire on Earth*. New York: Henry Holt, 1995.

Ricketts, Edward F. *Breaking Through: Essays, Journals, and Travelogues of Edward F. Ricketts*. Katharine Rodger, ed. Berkeley: University of California Press, 2006.

Ricketts, Edward F., Jack Calvin, and Joel W. Hedgpeth. *Between Pacific Tides*, 5th ed. David W. Phillips, ed. Stanford: Stanford University Press, 1985.

Robinson, W. W. *Land in California: The Story of Mission Lands, Ranchos, Squatters, Mining Claims, Railroad Grants, Land Scrip, Homesteads*. Chronicles of California. Berkeley: University of California Press, 1948.

Rodger, Katharine A. *Renaissance Man of Cannery Row: The Life and Letters of Edward F. Ricketts*. Tuscaloosa: University of Alabama Press, 2002.

Ross, Michael Elsohn, and Laurie A. Caple. *Flower Watching with Alice Eastwood*. Naturalist's Apprentice. Minneapolis: Carolrhoda, 1997.

Rothenberg, David. *Survival of the Beautiful: Art, Science, and Evolution*. New York: Bloomsbury, 2011.

———. *Why Birds Sing: A Journey into the Mystery of Bird Song*. New York: Basic, 2005.

Rothschild, Miriam. *Dear Lord Rothschild: Birds, Butterflies and History*. Glenside, PA: Balaban, 1983.

Russell, Bertrand. *The Scientific Outlook*. Glencoe, IL: Free Press, 1931.

Sagarin, Rafe, and Anibal Pauchard. *Observation and Ecology: Broadening the Scope of Science to Understand a Complex World*. Washington, DC, Island Press: 2012.

Schrepfer, Susan R. *The Fight to Save the Redwoods: A History of Environmental Reform, 1917–1978*. Madison: University of Wisconsin Press, 1983.

Sharpe, Grant W. *Interpreting the Environment*. New York: John Wiley, 1976.

Sharpe, Grant W., John C. Hendee, and Wenonah Sharpe. *Introduction to Forests and Renewable Resources*, 7th ed. New York: McGraw Hill, 2003.

Shillinglaw, Susan. *Carol and John Steinbeck: Portrait of a Marriage*. Western Literature Series. Reno: University of Nevada Press, 2013.

Simberloff, Daniel. *Invasive Species: What Everyone Needs to Know*. New York: Oxford University Press, 2013.

Sloan, Doris. *Geology of the San Francisco Bay Region*. California Natural History Guides. Berkeley: University of California Press, 2006.

Slobodkin, Lawrence B. *A Citizen's Guide to Ecology*. New York: Oxford University Press, 2003.

Slotkin, Richard. *Regeneration Through Violence: The Mythology of the American Frontier, 1600–1860*. Norman: University of Oklahoma Press, 1973.

Smith, James Payne, Jr. *Field Guide to Grasses of California*. California Natural History Guides. Berkeley: University of California Press, 2014.

Smith, Michael. *Pacific Visions: California Scientists and the Environment 1850–1915*. New Haven: Yale University Press, 1987.

Stallcup, Rich, Jules Evens, and Keith Hansen. *Field Guide to Birds of the Northern California Coast*. California Natural History Guides. Berkeley: University of California Press, 2014.

Stein, Barbara R. *On Her Own Terms: Annie Montague Alexander and the*

Rise of Science in the American West. Berkeley: University of California Press, 2001.

Stein, Mark. *How the States Got Their Shapes.* New York: HarperCollins, 2008.

Steinbeck, John. *Cannery Row.* New York: Viking, 1945.

———. *The Log from the Sea of Cortez.* New York: Viking, 1951.

———. *To a God Unknown.* New York: Penguin, 1933.

Steinbeck, John, and Ed Ricketts. *Sea of Cortez.* New York: Viking, 1941.

Steinhart, Peter. "Soaring with the Hawks," *California Wild* 50, no. 4 (fall 1997).

Tamm, Eric Enno. *Beyond the Outer Shores: The Untold Odyssey of Ed Ricketts, the Pioneering Ecologist Who Inspired John Steinbeck and Joseph Campbell.* New York: Four Walls Eight Windows, 2004.

van Dooren, Thom. *Vulture.* Animal. London: Reaktion, 2011.

Walljasper, Jay. *All That We Share: A Field Guide to the Commons.* New York: New Press, 2010.

Waterman, Talbot H. *Animal Navigation.* New York: Scientific American Library, 1989.

Weidensaul, Scott. *Living on the Wind: Across the Hemisphere with Migratory Birds.* New York: North Point, 1999.

Wheelwright, Jane Hollister, and Lynda Wheelwright Schmidt. *The Long Shore: A Psychological Experience of the Wilderness.* San Francisco: Sierra Club Nature and Natural Philosophy Library, 1991.

Wilson, Carol Green. *Alice Eastwood's Wonderland: The Adventures of a Botanist.* San Francisco: California Academy of Sciences, 1955.

Wilson, Edward O. *The Diversity of Life.* Cambridge, MA: Harvard University Press, 1992.

———, ed. *From So Simple a Beginning: The Four Great Books of Charles Darwin.* New York: W. W. Norton, 2006.

———. *Naturalist.* Washington, DC: Island Press, 1994.

Worster, Donald. *Nature's Economy: A History of Ecological Ideas*, 2nd ed. Cambridge: Cambridge University Press, 1994.

INDEX

mice, 81, 82
Michaux, André, 143–44
Midpeninsula Regional Open Space
District, 163–64, 167, 170
migrations, bird, 141–43, 378–81
Miller, Henry, 123
Millstein, Roberta, 231, 233–34, 235–36
Miner, Melissa, 21–22, 23
missions, Alta California, 113–15
Moby-Dick (Melville), 236, 237, 238
Monitor Change, 348–49
Moore, Earle K., 158, 159, 160
Moore, Frank, 160–61
Moore, Rebecca, 153–54, 156–57, 158, 159–60, 161–71, 179
Morrison, Kathleen, 121–22
movement of species, 84–85
Mrs. Dalloway (Woolf), 52
Muir, John, 123, 245, 294–95
myths, 7, 9, 390

naming, binomial system of, 41
National Phenology Network, 280–82
Natividad Island, 94–95
Nature Conservancy, The, 141–43
"nature needs half," 353–55
Nature's Notebook, 282, 284, 287
Neighbors Against Irresponsible
Logging, 168, 170
Nerds for Nature, 348–49
Newmark, William, 334–35
niches, 213–14
Noah's ark, 41–42
non-teleological thinking, 304–5, 319
North American Bird Phenology
Program, 337–38, 343
Notes on the State of Virginia (Jefferson),
8, 137–39, 145
nudibranchs, 29, 191, 227

Observation and Ecology (Sagarin &
Pauchard), 326–27
ocean acidification, 20–21, 33
ocean/coastal monitoring, 54–58, 60,
75–76, 77–78, 81–82, 95–96
ocean meltdown, 31–33
oil spills, 78–80, 226–27, 228
Okenia rosacea, 29, 30
otters, sea, 61, 62–63, 64, 65, 75
owls, burrowing, 81–82

Pacific Raptor Report, 362

Pacific Rocky Intertidal Monitoring
Program, 21–22
Packard, Heath, 94
Paine, Robert, 18, 102–3, 105, 106
Parker, J. J., 205, 210–11
parks, national, 99, 160–61, 214, 334–35
Parmesan, Camille, 367–68, 369–71
Parrish, Julia, 33, 80–81
Pauchard, Aníbal, 326–27
Peet's Coffee, 372–73, 374
petrels, ashy storm, 81–82
phalanx, 303
phenology, 278, 280–82, 337–38, 343
Philippines, 194
philosophy of science, 230–31
Pillar Point Harbor, 11–15, 16–17, 26–29,
391
Pimm, Stuart, 224–25
Plantago erecta, 268, 269, 271, 276–77
poetry, 312–13, 336, 337, 339
Point Blue Conservation Science, 31,
55, 57
Porter, Robert, 255, 256
Portolá, Gaspar de, 97–98
Prest, John, 40
Principles of Geology (Lyell), 44
Public Land Survey System, 149–50

Quiroste people, 98
Quiroste Valley, 112–13, 121–22

Raimondi, Peter, 17–18, 19–20, 21, 63,
299, 328–29
raptor banding, 381–85, 386
Raptors Are the Solution (RATS), 373,
374
Raven, Peter, 218, 269, 270–71, 272–73
Reckitt Benckiser, 373, 374
redwood trees, 167–70, 173–74, 175–80,
181–86
Redwood Watch, 178–79, 186
Refugio State Beach oil spill, 226–27,
228
Reynolds, Mark, 141–43
Ricketts, Ed
Campbell, Joseph, and, 292–93,
309–10
as citizen scientist, 328–29, 330
collection activities, 296–97
Darwin, Charles, influence of, 45
death, 321
education, 293–94, 295
followers, 306, 321, 322–23, 325

MARY ELLEN HANNIBAL is a longtime environmental journalist and the author, most recently, of *The Spine of the Continent: The Race to Save America's Last, Best Wilderness.* Her writing has appeared in *The New York Times, San Francisco Chronicle, Esquire,* and *Elle,* among many others. She is an Alicia Patterson Foundation Fellow, a recipient of the National Association of Science Writers' Science in Society Award, and a recipient of Stanford University's Knight-Risser Prize for Western Environmental Journalism. Currently a Stanford Media Fellow, she is a frequent speaker connecting the scientific community to the concerned public. She lives in San Francisco.